COSMIC ECOLOGY

GEORGE A. SEIELSTAD

COSMIC ECOLOGY

THE VIEW FROM THE OUTSIDE IN

UNIVERSITY OF CALIFORNIA PRESS
Berkeley *Los Angeles* *London*

University of California Press
Berkeley and Los Angeles, California

University of California Press, Ltd.
London, England

Library of Congress Cataloging in Publication Data

Seielstad, George A.
Cosmic ecology.

Includes index.
1. Cosmology. 2. Evolution. I. Title.
QB981.S445 1983 523.1 82-15944
ISBN 0-520-04753-2

Printed in the United States of America

1 2 3 4 5 6 7 8 9

To posterity, in the hope that there may be one; and especially to those immediate links which Dolores and I have forged with the forever after . . . Andrea, Carl, and Mark.

CONTENTS

PREFACE

The subject matter of this book is too varied and too extensive to lie within the expertise of any one person; nevertheless, the topics discussed are of importance to each of us. Each should wonder whence he came, realize how fortunate he is to have done so, and question why he is able to wonder at all.

There is, however, no use pondering whether this book should or should not have been written: the author simply had no choice. The material herein arose almost spontaneously from the confluence of profession with daily living. Two decades as a practicing astronomer at the California Institute of Technology's Owens Valley Radio Observatory have permitted me to live beneath the dominating presence of the Sierra Nevada. The natural beauty of the setting and the rich sample of Earth life it contains daily press themselves upon my consciousness. There is no escape (or any desire to do so) from consideration of the ecological relationships binding the setting to its inhabitants and the latter to one another. At the same time, my profession constantly reminds me of what a tiny sliver of totality commands my attention. Having tried to piece the two perspectives together in my own mind, I offer them here for others to consider.

The reader is hereby forewarned that science is always messy: every increment in knowledge raises more questions than it answers. Controversy is often the catalyst that initiates the incremental advances. But the reader will find little of that here. Although alternative theories abound at each step of the evolutionary progression presented, few will be specifically mentioned. To do so would risk obscuring the book's central thesis by marching the reader through a maze of side paths. I do, however, urge every reader to consult the bibliography at the end of each chapter for references to competing theories.

I apologize in advance to those who contributed the knowledge of which this book speaks. Few are mentioned by name. This accords with the author's belief that the nuggets of truth are the essence, not the individuals who discover them. Often discoveries are the result of the instruments available. Therefore one must be in the right place at the right time, a fortuitous circumstance as much as one resulting from exceptional insight. Besides, discoveries not made by one individual will eventually be made by another. In addition, few discoveries can be completely isolated from the scientific activity preceding them. To name any one person or group of people therefore offends those not mentioned without whose efforts the conditions for discovery would never have existed. None of this is intended to denigrate the contributions of many brilliant minds. Least of all is absence of attribution intended to imply that statements of fact are the products of the author's own research. Nothing could be more erroneous. I wish only to minimize obstacles that might interfere with the reader's concentration upon the central theme. Again, the references in the bibliography will lead the reader to the original sources of discovery.

The purpose of this book is to inform. If the same realization strikes the reader as it has the author, it will also alert. The author fervently hopes that awareness of the seriousness of our predicament will coerce us to find ways to survive it. Unflagging optimism convinces me we can find solutions to any problem to which we devote sufficient attention. No detailed solutions are presented, however, both because the author needs to leave some material for another book and because he feels obligated to elicit contributions to the meme pool (see chapter 10) from everyone. If none respond, the book is a failure.

Preliminary versions of the manuscript were read by Evelyn Eaton, Linda Goff, Tony and Carol Readhead, and David Seielstad. Their comments made the final product significantly better than it would otherwise have been.

Toni Bayer was more than an able typist. She also sharpened the focus in several places by querying its meaning.

Writing is an intensely introspective activity wherein ample rumination precedes the appearance of words on paper. During this stage, one's internal concentration is at the expense of external events. The author thanks especially his family, but also his colleagues, for suffering these spells good-naturedly.

A larger influence in shaping one's thoughts than the environment in which he lives is the people with whom he interacts. The person closest and therefore having the most effect on this manuscript is my wife Dolores.

George A. Seielstad

Bishop, California
May 1982

1
THE SCENARIO

For in fact what is man in nature?
A Nothing in comparison with the Infinite,
an All in comparison with the Nothing,
a mean between nothing and everything.

—PASCAL

Humans occupy a middle vantage point for studying nature. Compared with atoms we are immense; compared with the universe at large we are minute. But ironically, we are made of atoms, and we have descended in a natural way from the universe itself. What is more important, from this middle ground we can reach out, at least intellectually, in both directions to explore the realms of the very small and the very large. Moreover, it is crucial that we do so if we are ever to understand our role on this planet at this time. For we have inherited, from all the human generations that have preceded us on Earth, a cumulative body of knowledge so vast that we now have sufficient power to act in ways felt throughout the entire biosphere. These global powers have been acquired so rapidly and so recently that we as yet lack the wisdom to use them sensibly. We have therefore arrived at a fork in the road, a fork from which we cannot retreat. Either we consciously and collectively choose the path of harmonious existence within our environment, or we plunge recklessly down a dead-end road leading toward certain self-annihilation.

To view our predicament properly requires an enlightened and broadened perspective—indeed, a cosmic perspective. Without it our senses are deluged with so much detail that the broader patterns and rhythms of greater significance are obscured. Do we not all concentrate most on those features of our environment with which we are in direct and almost daily contact? And do we not fail to notice changes in those surroundings unless they occur quickly, that is, in time intervals that are short fractions of our own lifetimes? The danger with this merely human perspective is that it ignores a huge range of reality. We are, in fact, only parts of a system much older and much bigger than we. To comprehend that system, we must stretch our horizons in both space and time, thereby unshackling our restricted and distorted vision.

As a beginning step, consider our impression of Earth's magnitude. Imagine visiting, as an example, Death Valley in California. There one witnesses a scene of vastness, of immensity, of a land without limits—or so it seems. Indeed, it has been true historically that human intruders into this environment who underestimated its awesomeness were literally destroyed.

FIGURE 1
Death Valley National Monument. Ubehebe Crater dwarfs human intruders.

FIGURE 2
Death Valley from 570 miles above Earth's surface. Valley winds from upper left through picture's center. Scene from figure 1 is inconspicuous. (Landsat photo courtesy of NASA)

Contrast this with the view from a satellite. Now Death Valley seems but a minor scratch on the surface of the planet Earth. Since a valley that swallows us in its immensity is so insignificant a feature on our planet, we conclude that Earth must be an impressively large and substantial body. But we delude ourselves.

The depths of space may contain at least as many planets like Earth as there are grains of sand on all the beaches of the world; and yet none will be of any consequence in producing the faint smudges of light revealed in figure 3. Each smudge is a galaxy, and the entire collection is a cluster of galaxies. The light from each galaxy is the combined output of stars—hundreds of billions of them. Stars are so vastly bigger and brighter than planets that all the "earths" present in figure 3 contribute negligibly to the light there. We can safely conclude, therefore, that Earth itself is neither significant nor prominent cosmically.

FIGURE 3
Cluster of galaxies in Hydra. Bright spots are foreground stars in Milky Way Galaxy. Faint smudges are galaxies almost four billion light-years from Milky Way. (Palomar Observatory photo)

In probing the universe to the depths of figure 3, it matters not at all in which direction we gaze. Each yields the same overall impression. The universe has no preferred axis, no arrow defining a unique direction. Space is, in a word, isotropic. Nor does it matter how far we probe. A sample portion of the universe nearby has the same average properties as a remote portion, provided the two volumes are suitably large to include statistically significant numbers of objects. We may therefore describe the universe as homogeneous as well. These two properties, isotropy and homogeneity, describe a universe with no privileged locations, no exclusive neighborhoods. We (and other hypothetical beings elsewhere) must therefore erase all vestiges of a Ptolemaic, or geocentric, universe. Earth is not only insignificant but also arbitrarily situated.

Imagine the view of figure 3 is a glance homeward, acquired after an outbound journey at the speed of light lasting billions of years. How much matter would we have encountered during our long egress? The answer is almost none. To an excellent approximation the universe is empty. Regions between galaxies, and even between stars within galaxies, are better vacuums than man has ever been able to make in any physical laboratory on Earth. In fact, for every cube one inch on each side filled with stellar material, there are a thousand billion billion billion (10^{30}) such cubes that are empty. If all material contents were imagined smeared uniformly throughout the universe, then every four thousand trillion (4×10^{15}) cubic miles would contain scarcely an ounce. Astronomy could, with only slight exaggeration, be defined as the science that studies nothing. There is so much nothing there, however, that its sum can be quite substantial. What finer illustration of the vastness of space? Space is so extensive that sighting along any direction one eventually peers through considerable matter, despite its exceedingly sparse distribution.

Now consider returning homeward from the remote perch to which our imaginations have carried us. We shall travel, as on the outbound journey, with a passing light beam at its speed of 186,000 miles per second. We begin from a distance that took 4 billion years to reach. Initially, we bound billions

FIGURE 4
Andromeda Galaxy. Fellow member of Local Group of Galaxies, look-alike to Milky Way Galaxy, "parent" to two satellite galaxies, also shown. (Palomar Observatory photo)

of years at a leap. The cluster of galaxies toward which we aim comes into ever sharper focus, each member appearing bigger and brighter. When 140 million years from home, we notice that galaxies differ in form. After another 100 million years or so, we focus upon a minor cluster whose membership numbers only in the tens. It is the Local Group of Galaxies, among which is our own. Eventually, we arrive at the beautiful Andromeda Galaxy, a near

FIGURE 5
Earthrise over moon. The life that so distinguishes Earth occupies only a thin veneer over a base as desolate as the lunar surface. (Courtesy NASA/JSC)

twin to the Milky Way, our destination a scant 2 million years hence. At last we enter the Milky Way via a so-called spiral arm. Rarefied clouds of gas and dust occasionally interrupt the typical five-year interval between stellar encounters. Ahead looms the sun, its surface speckled with blemishes that, although minor, are each the size of Planet Earth. Eight minutes later we soar past a cold, dead rock with a fiercely pounded surface, the moon. Scarcely more than a second remains of our tremendous journey.

Where, exactly, will we arrive? At one of the least significant specks of rock and metal in the cosmos. Despite its cosmic irrelevancy, Earth is an object of exceptional beauty, a gleaming cosmic jewel contrasting stunningly with the black background of interplanetary space. It is also lonely, appearing magically suspended within a dark, yawning void. Likewise apparent is its fragility. Only the slimmest enveloping membrane of air and water shelters life from the harshness that surrounds it, a membrane whose relative dimension is but that of the shell on an egg. Finally, our view from the outside in reveals Planet Earth as a single entity, a unit. When looked at from this cosmic perspective, it is inconceivable that individual microenvironments, or the organisms within them, can function separately from the whole.

A proper understanding of our temporal horizons, as with our spatial, demands a mental emancipation. Each of us experiences a biological time: a cycle of birth, growth, and death. To this we add a sense of social time, an interval spanning four or perhaps five human generations, because that is the maximum number that simultaneously exist. Still earlier events fade rapidly into a historical time and carry sharply reduced emotional impact. But all these time scales are relatively infinitesimal in coping with the natural flux of cosmic history. We therefore suffer from a "temporal myopia." For example, one of the qualities often ascribed to the sky is stability. Certain cyclic events, such as the phases of the moon or the appearances of Halley's Comet, do indeed repeat faithfully and exactly, providing a soothing dependability. As a result, when contemplating the sky, people often sense feelings of eternity and immutability. In reality, however, the universe is a constantly changing, dynamic, evolving system. Today is different from yesterday. Tomorrow will be different from today.

To the best of our present knowledge the entire system erupted from a single point in space at a single instant of time less than twenty billion years ago. This event is graphically referred to as the "Big Bang." It is fruitless to search for its precise point of origin, for we are immersed in it. At the moment of birth, the universe was filled with its initial endowment of matter and energy. Although subsequently either could transform into the other, the sum of matter and energy would never change. Nor would those contents ever fail to fill the universe, since they simply have no place else to go. So we today are surrounded by the remnants of our origin.

The explosion that began our universe also launched it on a course of outward expansion. As a result, the universe constantly cooled. If it indeed started from a furiously hot explosion, it should still be filled with heat, although at a temperature reduced dramatically as a result of nearly twenty billion years of growth. This fossil form of heat has indeed been detected by the telltale radio signal it emits, providing decisive confirmation of a hot beginning.

Meanwhile, what of the original matter? Nearly all was hydrogen, the simplest and lightest of the chemical elements. In the early moments of the universe, however, both the density and the temperature of the hydrogen were extraordinary. Because of the high temperature, the particles moved very rapidly; because of the high density, they collided often. Occasionally hydrogen atoms collided so violently that they stuck together, forging the next heaviest element. Within minutes after the birth of the universe, one-tenth of the matter had in this way become helium, the rest remaining as hydrogen.* But already the universal expansion had carried individual particles to such great separations and had cooled, or slowed, them so drastically that the possibility of further fusion ended abruptly. For a very long time thereafter, the only chemistry possible was that involving the two lightest and simplest atoms.

---

*These numbers give the fractions of the total number of particles contributed by each element. Since helium is four times as massive as hydrogen, the approximate ratios by mass are hydrogen:helium = 3:1.

Given the chaotic nature of the universe's origin, it is unlikely that all the matter could have been distributed perfectly uniformly. Rather, there were probably some regions of excess density and others relatively rarefied. Extra-dense regions possess a stronger gravitational influence. They therefore add to their bulk by attracting neighboring particles. Eventually the collections' masses become so excessive they cannot resist self-collapse. They thereupon coalesce into galaxies. This formation stage for galaxies, including the Milky Way, required perhaps a billion years and took place at least ten to fifteen billion years ago.

The interior of a collapsing particulate collection as large as a galaxy could not be homogeneous either. Atoms within would not therefore be tugged toward any single location. Rather, several centers, distinguished by their above-average densities, would serve as focuses for fragmentation and subsequent collapse. As in the previous step within the universe at large, the oversized regions would grow ever bulkier until they, too, caved in under the burden of their own weight. In this way stars appeared, a process that started as soon as the galaxies themselves formed and that continues to the present.

It is unlikely that the process of gravitational collapse was so perfectly efficient as to consume every particle of matter in the vicinity. Instead, some dregs were left over, and these could have quickly been swept up into planets. Debris remaining after this stage in turn formed rings or planetary satellites or both. In this fashion, the sun and the accompanying solar system came into existence about four and a half billion years ago. The history of our home planet therefore occupies only the most recent fourth of the totality of cosmic history.

As the evolution of the universe has been traced in these pages so far, a hot, dense beginning from total chaos has given way to an expanding, cooling space within which immense configurations—namely, galaxies and stars—have materialized. The only atomic elements present, however, are hydrogen and helium in the rough numeric ratio of nine to one. Yet we know that today's Earth is blessed with an epicurean selection of chemicals with which nature can concoct endless recipes for new compounds. Where did all the other ingredients, from actinium to zirconium, come from? The answer is that they were forged in crucibles at the centers of now-dead stars.

Stars mimic living systems. They are born, live to maturity at metabolic rates determined by their masses, and die, spewing forth the matter from which their stellar offspring can take form. Throughout, they convert the light atoms of their birth into the heavier ones dispersed at death. They therefore constantly enrich the chemical inventory of their environment. The chemicals that constitute our beings were manufactured in the bowels of stars that today exist only as memories. Had the solar system formed earlier, before these ancestral stars had expended their life cycles, its constitution would have been too chemically restricted to have fostered our existence.

Of course, the evolution of matter did not stop at the simple production of raw chemical elements. The next step in the organizational structure of matter, the formation of molecules, was also taken. Some atoms have a natural electrical affinity for others, at least if they can be brought into close proximity. This suggests that a likely place to seek molecules might be in the same regions where stars are born, because these regions are known to be exceptionally dense. Indeed, the expectation has been amply fulfilled: more than fifty molecules have been identified, primarily by the characteristic signatures of their radio emission, in a list (table 1) that is still being expanded. Since star formation is a common occurrence in both our galaxy and its neighbors, the presence of molecules is not rare happenstance; they are ubiquitous. Nor are the molecules on Earth different from their interstellar analogues. The same chemistry operates in the Andromeda Galaxy as in the laboratory of your local university.

The individual molecular species are themselves of considerable interest. Who, for example, except perhaps in his wildest fantasies, would have foreseen the common occurrence of water, the most precious ingredient of all

TABLE 1 Interstellar Molecules

Number of atoms	*Inorganic*		*Organic*	
2	H_2	hydrogen	CH, CH^+	
	OH		CN cyanogen	
	SiO	silicon monoxide	CO carbon monoxide, CO^+	
			CS	
	NS			
	SO			
	SiS			
	NO			
3	H_2O	water	HCN hydrogen cyanide	
	HN_2^+		HCO, HCO^+	
	H_2S	hydrogen sulfide	HNC	
	SO_2	sulfur dioxide	CCH	
	HNO		OCS	
			HCS^+	
4	NH_3	ammonia	HNCO	H_2CO formaldehyde
			C_3N	H_2CS thioformaldehyde
			HCNS	
5			CH_2CO	H_2CNH
			C_4H	HCOOH formic acid
			H_2NCN	HC_3N
6			CH_3CH	CH_3OH methyl alcohol
			$HCONH_2$	CH_3SH
7			CH_3NH_2	C_2H_3CN
			CH_3C_2H	HC_5N
			CH_3CHO	
8			CH_3COOH	
9			C_2H_5CN	CH_3OCH_3 dimethyl ether
			HC_7N	CH_3CH_2OH ethyl alcohol
11			HC_9N	

H = hydrogen, C = carbon, N = nitrogen, O = oxygen, Si = silicon, S = sulfur earthly life, in the spaces between stars? Or, for that matter, of formic acid, the substance that gives a bee sting its wallop? Or of a molecule as chemically complex as alcohol? Especially significant are the many cosmic molecules containing carbon, an atom so open to bonded partnerships that it forms the backbone of the long molecular chains of which all flora and fauna on our planet are constituted. The fundamental links of these chains, called amino acids, are shared by humans with bacteria, birds, bats, bees, and bromeliads, among others. And amino acids are natural end products of laboratory experiments simulating conditions believed to have prevailed on primitive Earth. In a famous experiment, Stanley Miller discharged a spark of electricity into a container filled with molecular hydrogen, water, methane, and ammonia. Various amino acids resulted, including some of the ones from which all Earth life is constructed. The input molecules were representative of those present in the Earth's original atmosphere, while the electrical spark could have been lightning. We now know that similar molecules exist in abundance at diverse locations in the universe. The molecules from which entire communities of living systems can be nurtured abound! The experiment in living which it has been Earth's immense good fortune to have hosted need not be unique. Nevertheless, we shall focus our attention for a moment on the blossoming of life in just this one locale.

The atomic, then chemical, evolution that preceded the formation of the solar system paved the way for the emergence and subsequent development of life on (at least) one planet. But a key question is how the relatively simple compounds initially present were converted to the lengthy chains that are the essence of the living process today. As we have seen, amino-acid formation was a natural consequence of the geologic and chemical conditions prevalent early in the planet's history. If, as the young planet cooled and solidified, the steam that vented from its fissures cooled and condensed, then vast oceans were created. These formed perfect reservoirs for the collection and concentration (perhaps in tidepools) of the molecular varieties being spontaneously produced. Within the resultant broth, innumerable trial-and-error experiments inevitably took place in which different chemical combinations tested their abilities to persist despite the challenges of their environment. The most suitable survived in greatest numbers. An obvious example of suitability was the ability to reproduce. Any molecule that could assemble an exact copy of itself from prevalent raw materials would soon outnumber all those whose appearance resulted purely from chance. Furthermore, in any struggle with a hostile environment a buffer, or wall, moderating the extremes of variability is obviously an immense advantage—an advantage seized with the advent of cells more than three billion years ago.

This type of gambler's game, in which oddball systems (mutations) adventitiously appear and try to "strike it rich" by beating out their competitors, rages on and on, even today. The arbiter of who wins and who loses is nature itself, and evolution is the device by which nature selects those systems that will produce successful progeny. Over time it builds complexity from simplicity. Hummingbirds, for example, appear long after jellyfish. Though we need not, perhaps cannot, retrace every step in the progression, we believe that living matter sprang from the nonliving; that having done so once, all life derived from other life; and that the process could be universal, not unique or peculiar only to this planet. The essentials—some precursory molecules, an energy input, a suitable environment, and ample time—are available at nearly countless locations throughout the universe.

From our newly acquired cosmic perspective we can now see that life is a continuum traceable to the creation of space and initiation of time at the Big Bang itself. We are swept up in a natural cosmic flow—along a rivulet branching off a stream forking off a river, and so on, back to and beyond the fountain from which all life sprang. Whereas each of us lives only momentarily, a pattern and structure persists in our descendants, stamping them uniquely as humans. As a whirlpool persists despite the constant changing of its water, so we can view our existence in the human lifestream. Although we are therefore individually insignificant and transient participants in an evolving system vastly transcending us, as a species we are necessary links between the aeons of the cosmic past and all that is yet to follow. Our salvation and immortality lie in the human lifestream, which cannot be divorced from the stream of events off which we branched. Preservation of the flow transcends our individual significance.

Humans appear late on Planet Earth. We can measure our history in millions of years, whereas billions are needed to encompass that of Earth itself. We have been present a mere tenth of a percent of the history of this planet. Despite entering the game of life late, with a rich, diverse, and thriving community of living systems preceding us, we have been neither bashful about making our presence felt nor humble in evaluating our significance. As a matter of fact, we rank ourselves at the very pinnacle of all living matter. Have we any justification? Perhaps, and if so it lies in our ability to communicate—not merely with contemporaries, for we know that whales and dolphins, among others, have mastered that form of communication. We alone, however, can also communicate across generations. Messages are transmitted to us down through the ages: verbally as well as through the written word, the pictorial word, and the recorded word, and through imagery encoded in our art and music. We therefore need not learn every lesson anew. To the knowledge we inherit, we can only add. The result is a rapid accumulation, an exponential explosion, of both knowledge and the power it conveys. Consequently the pace of our cultural evolution is dramatically accelerated over that of the atomic, chemical, and biological evolutions that preceded it.

Consider flight, for the sake of illustration. Birds, bats, and bees became fliers only by evolving genetically over millions of years. Man, by contrast, became the fastest and most powerful flier of all, not by reconstructing his genotype but by building machines to help him become airborne. In doing so, he compressed the time required by factors of thousands!

The astonishing rapidity of this and other changes is a consequence of our greatest evolutionary advantage, intelligence. But let us not be parochial. Wherever in the universe life evolves, it undoubtedly follows a like progression from the simple to the complex, threading its way precariously through the screening process of natural selection. In coping with the changing conditions of any environment, an obvious, therefore perhaps inevitable, advantage accrues to an organism equipped with some sensory apparatus for mapping out the environmental parameters. Then a central processor for handling the flood of incoming data becomes valuable. A storage bank, or memory unit, for storing consequences of different reactions to external stimuli seems another aid. Finally, a decision-making capacity for selecting among possible options becomes equally beneficial. May we not at least speculate that this sequence leads toward intelligence? And that it has a certain inevitability wherever evolving living systems interact with a changing environment?

Can we now better understand Earth's predicament at this precise moment? Recall that interstellar molecules, closely related to the organic ones composing all Earth life, are widely distributed. Furthermore, there are almost certainly ample locations—perhaps on planets orbiting other suitable sustaining stars—where their apparently easy production can be fostered. If they eventually succeed in making self-replicating systems, then life may follow. The resultant living systems will tend, with time, toward greater complexity and higher intelligence. Eventually at least one species of the community may acquire sufficient intelligence to permit accumulation of knowledge at an exponentially growing rate. With that knowledge comes power, the growth in the former driving that in the latter. Finally, almost instantly, the power at the command of the dominant species becomes so great that it can affect events on a planetwide scale. This moment of acquisition is unique for that particular planet. It is also exceedingly dangerous, for the dominant species has no experience in managing its awesome abilities.

Thus the Earth stands precariously poised today, with we humans the species that has precipitated the crisis. It is worth reemphasizing that our problems are different in nature from those encountered previously. We now encounter definite limits. In just the last century the speed of communication has accelerated from that of the pony express to that of light; but we shall never transmit messages any faster. Likewise our mode of travel has changed from clipper ships to spaceships; but we shall never circumnavigate the globe at greater than orbital speed. The sheer number of human inhabitants has nearly tripled in the same hundred years; but we cannot emigrate into space fast enough to halt the rise. We have consumed much of the world's oil; but we dare not burn so much more of it, or of its hydrocarbon relatives, that the global temperature rises even a few degrees. Our military encounters have changed from cavalry charges to nuclear holocausts; but we cannot kill deader than dead. These, and other, recently encountered limits portend a time of momentous decision.

How are we to confront our challenge? We must first recognize it, acknowledge its seriousness, and then plot our course of action. It is not a time for the timid or for the hesitant. An entirely new type of evolution is needed, which I call collective, participatory evolution. It is collective because it involves *societal* decisions: not decisions made by only the rich nations, or for that matter by only the poor; not decisions made solely by one politico-economic system or another; but decisions that involve and consider the entire worldwide community of human beings. The new evolution is participatory because it cannot be passive. Failure to make decisions is in itself a decision—of the most reckless and irresponsible kind. Natural selection is the rule that governed all previous evolution: atomic, chemical, and biological. Things and events just happened. But collective, participatory evolution demands conscious selection followed by decisive action.

We humans alone on this planet can override the genetic instructions of our inheritance and the indoctrination of our cultural upbringing. We alone can project alternative futures in our imaginations. If we are both bold and wise in choosing among them, we can assure a future so glorious that all preceding history will seem a pale prelude to it. If, however, we are timid and foolish, we will forfeit an opportunity that comes but once to each living community on every suitable planet. The choice belongs to us, the present generation of humans on Planet Earth.

FOR FURTHER READING

CHAISSON, ERIC J. 1981. *Cosmic dawn*. Boston: Little, Brown and Company. 302 pp.

DAWKINS, RICHARD. 1976. *The selfish gene*. New York: Oxford University Press. Pp. 13–21.

GURIN, JOEL. 1980 July/Aug. In the beginning. *Science 80* 1(5):44–51.

LINDBERGH, CHARLES A. 1979. Letter to Emilio Q. Daddario. *Science* 204:1392–1393.

MURDY, W. H. 1975. Anthropocentrism: a modern version. *Science* 187:1168–1172.

OLSON, R. L. 1975. Response to Murdy article. *Science* 189:594.

SEIELSTAD, GEORGE A. 1978 Nov./Dec. Cosmic ecology: the view from the outside in. *Mercury* 7(6):119–124.

2
OUR PLACE IN SPACE

His homocentric view was undeterred by inhabiting a minor galaxy, far from its center, sustained by a very ordinary star on an insignificant planet on which he was a most recent expression. . . . He was, in fact, pre-Copernican in believing that man bestrode the earth around which the sun revolved, in turn the center of the surrounding cosmos, testifying to the primacy of man.

—Ian L. McHarg

That his vision has plumbed the depths of space represents one of man's greatest intellectual achievements. Having confined our travels thus far to the few light-*seconds* immediately adjacent to Earth, we can nevertheless project our imaginations over distances measured in tens of billions (10^{10}) of light-*years*. To do so we work from near to far, committing first one gap to our comprehension, then using it as a reference rod for spanning still larger chasms. The result is a rather shaky structure whose every step becomes more uncertain. The methods, though, are sound, and more and better measurements will constantly refine our precision.

The practicing astronomer intercepts signals—be they radio waves, light, X rays, or whatever—which all travel at the same speed, approximately 186,000 miles per second. To him, then, this speed provides a natural standard for gauging distance: the interval between transmission and reception of a signal, multiplied by the speed above, converts to a distance. The light-year, being the product of 186,000 miles per second with the thirty-two million seconds in a year, is therefore nearly six trillion (6×10^{12}) miles. This unit neatly unifies space and time, for a distant object is seen not as it is now but as it was then, when the signal revealing its presence began its journey. The farther we probe in space, the earlier we explore in time. Geometry and history are in this sense intertwined.

Earth's diameter can be traversed by a light ray in one twenty-fifth of a second; light can circumnavigate our globe about seven and one-half times in a single second. Earth's nearest cosmic neighbor, the moon, is smaller still: a hollow Earth could harbor almost fifty moons. Some readers will recall, during man's exploration of this satellite, viewing the tense moments when lift-off for home was scheduled. The countdown from Houston—"ten, nine, eight, . . ."—reached our homes nearly instantaneously, so at the count of "zero" it was distressing to see nothing happen. The explanation, of course, is that the astronaut received his count only after a delay of one and a third seconds, the light-travel time between planet and satellite. An equal delay

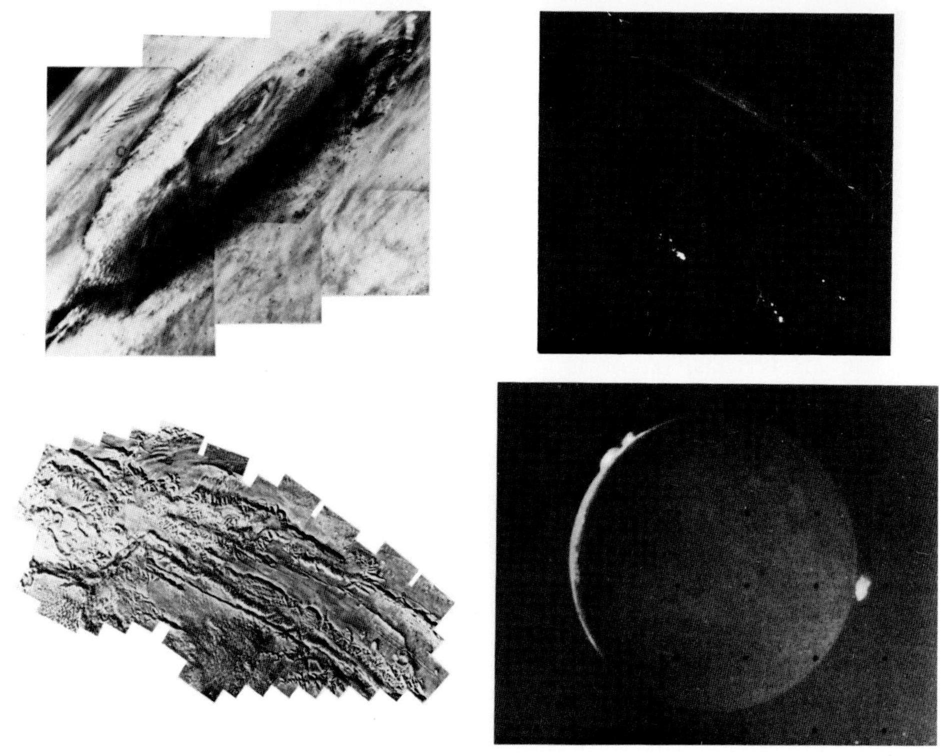

FIGURE 6
Solar system marvels. Upper left, Olympus Mons, a Martian volcano 15 miles high and 360 miles across at its base. Lower left, Valles Marineris, largest rift valley in solar system, 2,500 miles long and 75 miles wide. Upper right, night half of Jupiter, revealing several lightning flashes and an aurora (faint curve at top). Lower right, three simultaneously erupting volcanoes on Jupiter's moon Io. (All photos courtesy NASA/JPL)

preceded our witnessing of his response. The pause was long enough to evoke concern and demonstrated light's finite speed of propagation.

Only light-minutes to light-hours beyond this physically explored region lie eight other planets. Among these are the colossus, Jupiter, within whose interior thirteen hundred earths could be crammed, and diminutive Pluto, scarcely the equal of our moon. Earth is precisely mid size, with four planets larger and four smaller. Furthermore, our planet's most arresting features are not unsurpassed: the majesty of Mount Everest is dwarfed by Mars' Olympus Mons, as is the Grand Canyon by Valles Marineris. Jupiter's aurora outshines our own; the volcanoes on its satellite Io are more explosive than Etna, Vesuvius, or Krakatoa. And though we yield to none in appreciating the mysterious, haunting presence of our moon, in quantity we are decidedly inferior. Whereas we are faithfully "moonogamous," some other planets enjoy more than a dozen satellites, and the average is nearly five.

The cosmically local resident whose supremacy is unchallenged is the sun. All but one seven-hundredth of the solar system's mass is concentrated within it. The gravitational pull attendant to its bulk binds all nine planets in their orderly paths through the heavens. Its diameter exceeds Earth's by more than a hundred times; in volume the ratio exceeds a million. Earth, in fact, is comparable in size to one of the minor blemishes, or spots, that periodically mar the solar surface. The sun's fierce internal furnace often spits matter to heights in excess of the Earth–moon separation. The light, heat, and other radiant energy from this furnace reach Earth after some eight minutes, nourishing and sustaining the life there. Only after another five and a fraction hours will they pass Pluto on their way to the fringe of the solar system. There, at approximately a light-year, they encounter a swarm of frozen cometary debris.

Before exploring further, we must see the sun not as do creatures whose existence is wholly dependent upon it but as would others whose breadth of vision excludes parochialism. These others see mediocrity, for despite its supremely critical importance for life on Earth, among the immense family of stars the sun is only typical. In terms of mass, size, temperature, chemical composition, power output, and countless other stellar variables, our generous benefactor is merely average. Even in location within the stellar family comprising our galaxy, the sun resides with no distinction: it sits twice as far from the galaxy's center as from its edge.

Let us resurvey the extent of our system, beginning with the one unit whose magnitude we have experienced, at least vicariously, via the Apollo astronauts. Lay that Earth–moon span end to end sixty thousand times to

FIGURE 7
Solar prominence. Matter suspended above surface by magnetic fields to heights comparable to Earth–moon separation. (Big Bear Solar Observatory, California Institute of Technology)

reach the last material trapped within the sun's gravitational grip. So scales the solar system, the first organized unit in our outward calibration of spatial extent.

How far the stars? The nearest to our own is the whirling triplet, Alpha Centauri, four and a third light-years hence. This separation, four to five times beyond the influence of an individual star, is fairly typical of interstellar space. Were we to measure it in solar diameters, we would discover thirty million suns strung side by side, like beads on a necklace, between nearest neighbors. The galaxy is, it would seem, extremely porous. Of course, not all stars exist in splendid isolation. The sun, in fact, is distinctly idiosyncratic in this respect. More common are associations or clusters of stars, some loose or open, others dense and globular.

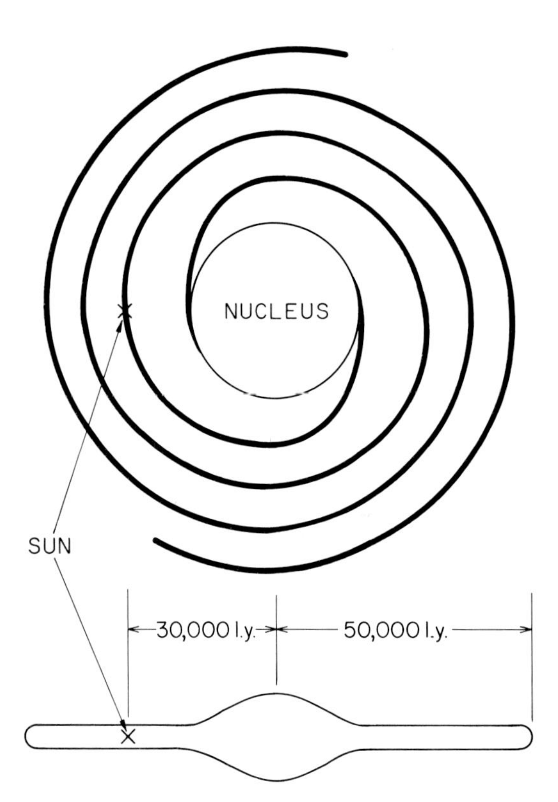

FIGURE 8
Schematic diagram of Milky Way Galaxy and sun's location within it.

The stars of our galaxy are not scattered randomly. The brightest and youngest, in particular, are nearly all confined to a thin plane or disk, within which they congregate in winding filaments, or arms, spiraling outward from the center. The so-called Local Arm, the one containing the sun, resembles a tube a thousand and a half light-years in diameter. A trough of sparser stellar density, wide enough for four such tubes, separates adjacent arms. The entire disk of our Milky Way Galaxy, when viewed face on, is a hundred thousand light-years in diameter, nearly twenty thousand times the typical distance between stellar neighbors. Within this ponderously turning disk are hundreds of billions of stars, many at least as luminous as the sun. And even though we have expanded our horizons beyond the solar system by a full hundred thousand (10^5) times, we have reached only the edge of our immediate stellar environs. Beyond lies a universe of galaxies as numerous as are the stars within our own.

The nearest to the Milky Way are the twin Clouds of Magellan, celestial treats for southern observers even using only the unaided eye. Their distance is scarcely twice the Milky Way's diameter; the three galaxies, in fact, constitute a trio connected by a gaseous bridge. But the Milky Way is clearly dominant, its smaller neighbors having barely a third its extent. The first real extragalactic equal is the Andromeda Galaxy (fig. 4), look-alike to our own. At more than two million light-years, twenty Milky Way diameters, it is the farthest object man can see without a telescope; equivalently, of course, it sends the oldest signals our unaided eyes detect. It, too, has companions, some almost immediately adjacent and others somewhat farther. Together the Milky Way and Andromeda anchor twenty or so companions in a bicentric cluster, called the Local Group, whose spatial extent reaches one and a half times the separation between the focuses.

Perhaps one universal characteristic has already become apparent: the gregariousness of celestial objects. Moons accompany planets, which are retained by stars. Stars gather in pairs, triplets, or other associations up to hundreds of thousands. These stellar aggregations are bound into galaxies, which themselves cluster. It seems that the most likely place to detect a celestial object is near another object. The hierarchy does not end yet, however. Clusters of galaxies are not rare. They abound, in equal numbers in all directions and at seemingly all distances. Some have only handfuls of members, but others contain thousands. The Virgo Cluster, named for the constellation through which it is seen, is an example of an immense configuration. With a membership fifty times that of the Local Group, but confined in thirty times the latter's volume, Virgo lies fifty million light-years hence. Not far beyond, galaxies temporarily become sparser. This suggests that all the clusters within some considerable distance from us are conjoined into a supercluster whose diameter of two hundred million light-years is roughly seventy times the diameter of the Local Group! As far as we know, superclusters are the largest organized structures of matter in the universe.

The boundary of the Local Supercluster, however, is not the boundary of

space. Galaxies, their clusters, and superclusters extend in every direction as far as technology has permitted us to peer—to distances measured in tens of billions (10^{10}) of light-years. Our temporal horizons recede to equivalently remote epochs. Can we imagine such vastness and seeming eternity? Scarcely, but perhaps analogies to similarly enormous quantities assist. For example, the distance to the farthest galaxies, as measured in Milky Way diameters, is the same ratio of a hundred thousand (10^5) that relates the Milky Way's diameter to the solar system's. And both the number of galaxies within our purview and the number of stars in our galaxy total in the hundreds of billions (10^{11}).

Perhaps now we can endure an experience at once humbling and ennobling, a look at the nearest million galaxies. No wider map has yet been produced. Were our eyes more sensitive by 160,000 times, we would see our meganeighbors scattered in the tangled, knotty pattern of figure 9. If galaxies had no tendency toward association but instead were located at, say, lattice points of a uniform crystal, the map would be shaded an even gray. The true distribution, while irregular, is nevertheless uniform: the fibrous cotton-candy appearance averaged over one general direction matches that over every other. Can we examine so grand a view without imagining fellow observers peering back, likewise glorifying this cosmic architectural majesty?

Galaxies continue on and on, challenging our instruments with the faintness of their messages. But do they extend without limit? If so, and if they have been radiating light forever, and if they are stationary, then every line of sight will eventually intercept a galaxy. There will be no dark patches of sky. Instead the sky will everywhere be as bright as a typical galaxy; yet we know it

FIGURE 9
A million galaxies within approximately the nearest 1.4 billion light-years. Are fellow observers peering back? (Courtesy M. Seldner, B. Siebers, E. J. Groth, and P. J. E. Peebles, Princeton University)

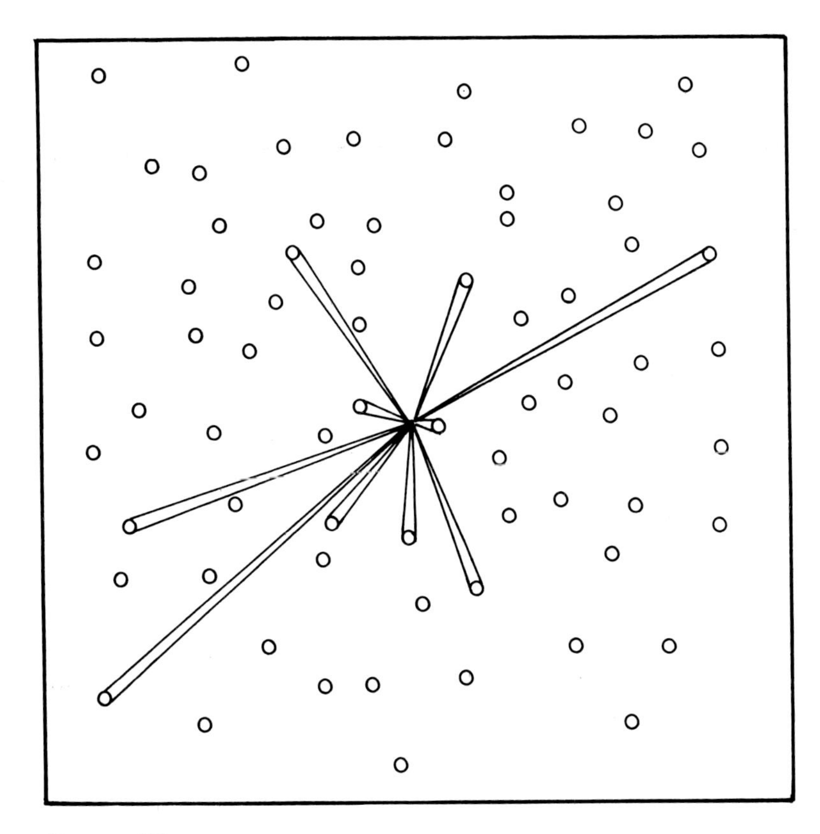

FIGURE 10
Lines of sight intercepting stationary galaxies. If galaxies extended forever in space and had radiated for all time, all lines of sight would end in a galaxy. The sky would have no dark patches.

is not, a paradox named after the German astronomer Heinrich Olbers (1758–1840). Apparently our fundamental assumptions are in error. Today we do not believe that galaxies have been shining forever, or that they are fixed. Rather, we believe they all formed some finite time ago and that all recede from one another as part of a universal expansion of space. There are accordingly lines of sight that penetrate to distances so great they are looking back in time to an era before galaxies were luminous. Furthermore, the apparent intensity of light from the most distant galaxies that are seen is diminished by the rapidity with which they flee us.

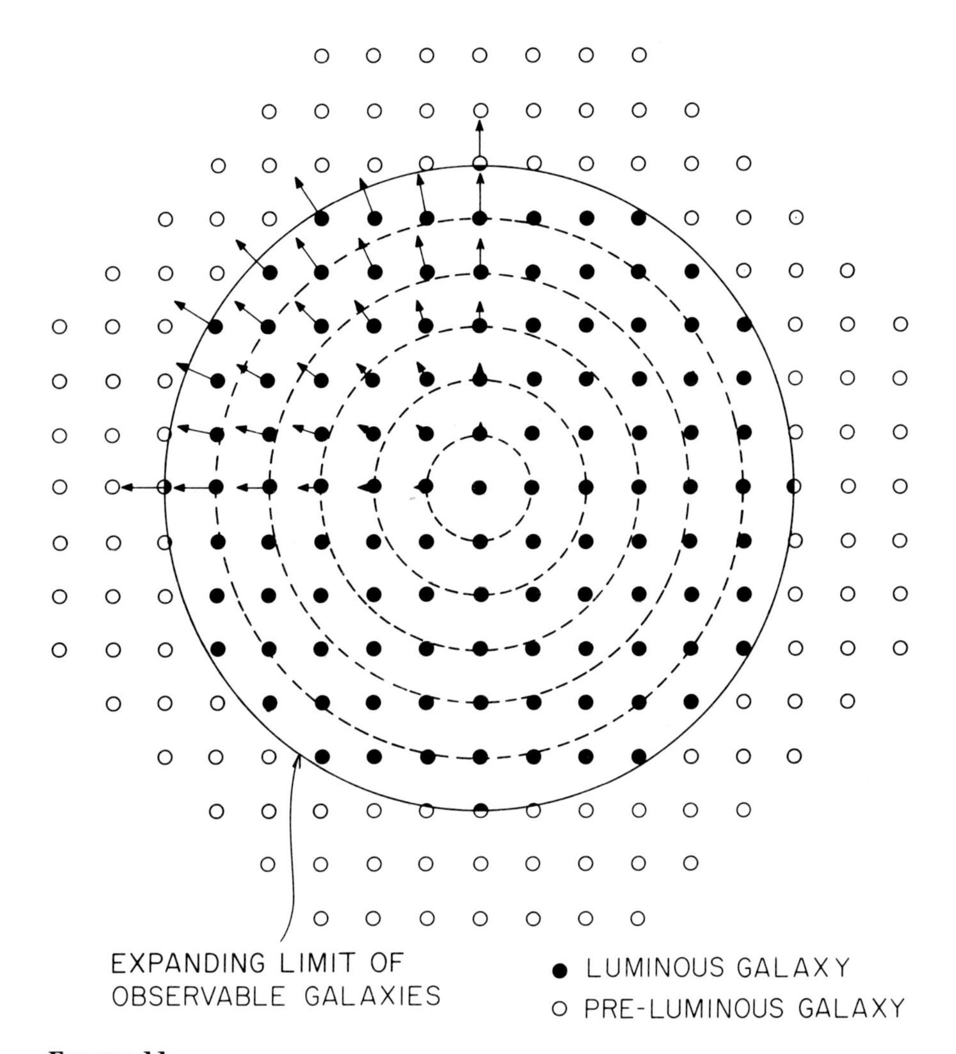

FIGURE 11
Olbers's paradox. If galaxies have been radiating for a finite time only, there will be a boundary (solid circle) beyond which no galaxies are seen. Some lines of sight consequently miss all luminous galaxies and have nothing to intercept beyond. Accordingly, there are dark patches in the sky. Furthermore, an expanding universe (indicated for one quadrant only) carries galaxies away at velocities proportional to their distances. Distant spherical shells thereby lose more galaxies than they gain. Successively more distant shells contribute progressively less light to the central observer. The sky can never achieve infinite brightness.

TABLE 2 **Cosmic Address**

Locale	*Factor larger than preceding step*
Earth	–
Solar system	8,000,000,000
Local Spiral Arm	1,500
Milky Way Galaxy	70
Local Group of Galaxies	30
Local Supercluster	70
Universe	100
Product of all factors	$2 \times 10^{20} = \frac{\text{size of universe}}{\text{size of earth}}$

It is natural, in the face of evidence for expansion, to ask from what center it proceeds. The question, however, ignores the possibility that the space of the univese may be warped or curved. If so, there is no single center. Instead, all points are equally central. To comprehend how so democratic a situation can exist, consider a two-dimensional analogy, the surface of a sphere. On our globe no single point—not Washington or Moscow or, for that matter, Addis Ababa—can be considered the geometric center of the spherical surface; more correctly, all can be considered centers. Only a short mental extrapolation is necessary to visualize the universe as a curved volume whose every interior point is as central as every other.

Doubtless it is difficult to grasp the vastness of the universe, the immensity of the yawning chasm in which we are immersed. Nevertheless, our goal has been to develop an appropriate sense of place, a spatial orientation undistorted by our limiting human finitude. We can now see that we inhabit the merest shred of cosmic flotsam adrift in the oceans of space. Moreover, we orbit an ordinary star situated inconspicuously within a mediocre galaxy belonging to a minor cluster of galaxies cast among the multitudes constituting the Local Supercluster, a gathering that distinguishes itself not at all from countless like ensembles in the universe. In short, our home is a cosmic trifle.

FOR FURTHER READING

Beatty, J. Kelly; O'Leary, Brian; and Chaikin, Andrew, eds. 1981. *The new solar system*. Cambridge: Cambridge University Press and Sky Publishing Corporation. 224 pp.

Boeke, Kees. 1957. *Cosmic view*. New York: The John Day Company. 48 pp.

Harrison, Edward R. 1981. *Cosmology*. Cambridge: Cambridge University Press. Pp. 249–265.

Hodge, Paul W. 1966. *Galaxies and cosmology*. New York: McGraw-Hill Book Company. 179 pp.

Sandage, Allan; Sandage, Mary; and Kristian, Jerome, eds. 1975. *Galaxies and the universe*. Chicago: University of Chicago Press. 818 pp.

Sciama, D. W. 1961. *The unity of the universe*. New York: Doubleday & Company, Inc. (Anchor Books). 213 pp.

Shipman, Harry L. 1976. *Black holes, quasars, and the universe*. Boston: Houghton Mifflin Company. Pp. 224–250.

Struve, Otto. 1959. *Elementary astronomy*. New York: Oxford University Press. Pp. 3–13.

3
OUR MOMENT IN TIME

Time is just Nature's way of making sure everything does not happen at once.

—ANON.

If the universe were truly static, time itself would be frozen. It is only through the process of change that we sense the passage of time. Physiological changes, for instance, accompany the aging of our bodies. The rotation of the planet Earth defines the day; its revolution around the sun, the year. A galactic year is defined by the time elapsed between reappearances of the sun at the same point in its orbit about the center of the Milky Way: it equals 250 million of our earthly years.

We are endeavoring, however, to elevate our vision beyond the merely local. From the grandest perspective, when galaxies themselves are treated as point masses, what phenomenon of change demonstrates the passage of time? What systematic action governs the ticking of a clock for the whole universe? The answer is expansion—not of the galaxies themselves but of the entire space in which they are immersed. Initially, an inflating universe might seem strange, but a static one would be stranger still. A system once at rest would collapse inward under the mutual pull of every galaxy for every other. The condition of rest would therefore necessarily have been preceded by an epoch of expansion. In either case, stasis would at best be a temporary option.

The evidence for an expanding universe begins with the observations that virtually all external galaxies appear to be fleeing from our own and that the greater their distances from us, the faster they recede. The relationship between speed of recession and distance is a strict proportionality at speeds much slower than light's, a doubling in one producing a doubling in the other. This is illustrated in figure 12, called a Hubble diagram in honor of the astronomer Edwin Hubble (1889–1953), who firmly established its validity. Determining how much the speed increases for each increment in distance has not been easy in practice. Today's best (though still imprecise) estimate is that the recessional speed increases by about ten miles per second for every jump in distance of a million light-years. That is, a galaxy at twenty million light-years moves from us one hundred miles per second faster than one at ten million light-years. The quantity relating velocity and distance, called Hubble's constant (H), is a crucial parameter of cosmology. Graphically, it is the slope of the line in figure 12.

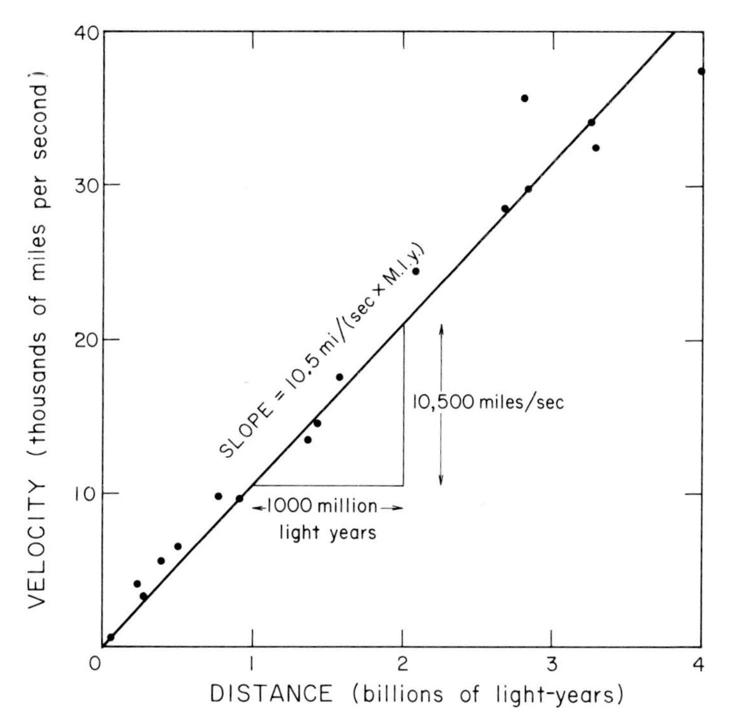

FIGURE 12
Hubble diagram. Recessional velocities of galaxies as functions of their distances. Straight line fits data at velocities much slower than light's. Line's slope, called Hubble's constant, yields distance estimates when velocities are known. Its inverse, the Hubble age, twenty billion years, is upper limit to age of universe.

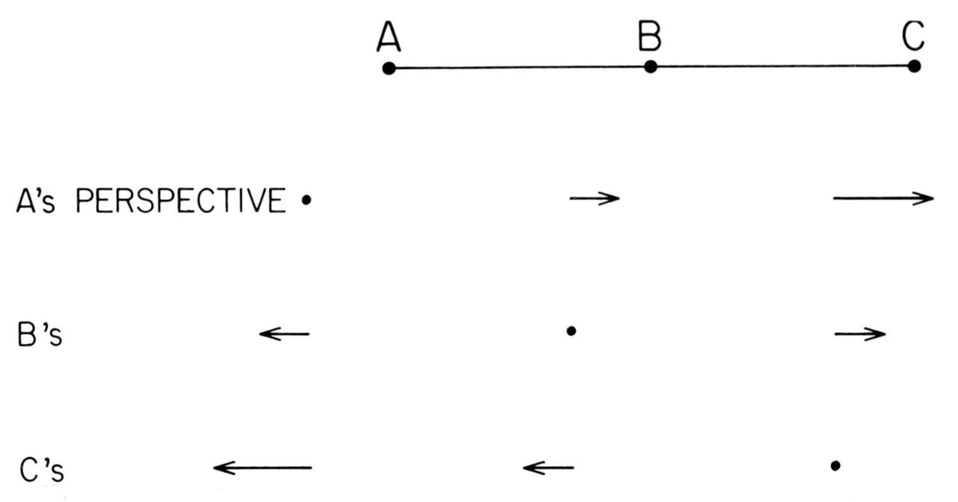

FIGURE 13
One-dimensional expanding universe. If universe has no preferred direction, central observer B must see A and C moving oppositely from him with equal speeds. Both A and C then see galaxy velocities exactly proportional to distance.

Why has the Hubble diagram been cited as evidence for an expanding universe? To see why, consider first a one-dimensional universe with just three collinear, equally spaced galaxies. We would prefer—indeed, extrapolating the Copernican viewpoint beyond the solar system, we would expect—that at any given time the universe would appear the same to observers on any galaxy looking in any direction. An observer on the central galaxy who sees one galaxy fleeing from him must therefore see the other moving oppositely with the same speed, else his space would have a preferred direction. What, then, is the perception from either extreme galaxy? Observers on them see a nearest neighbor moving away with one velocity and a second neighbor at twice the distance moving with twice that velocity. In other words, the symmetry present initially will be preserved forever after if, and only if, velocity is proportional to distance.

Of course, the example presented is far too restrictive. Imagine instead a two-dimensional universe that is precisely homogeneous initially, each galaxy occupying one lattice point of a square array. An arbitrarily situated observer who views an isotropic expansion, each galaxy appearing to flee from him with a velocity proportional to its distance, will note that his universe maintains its identical homogeneity at all times. So too, however, will every other observer.

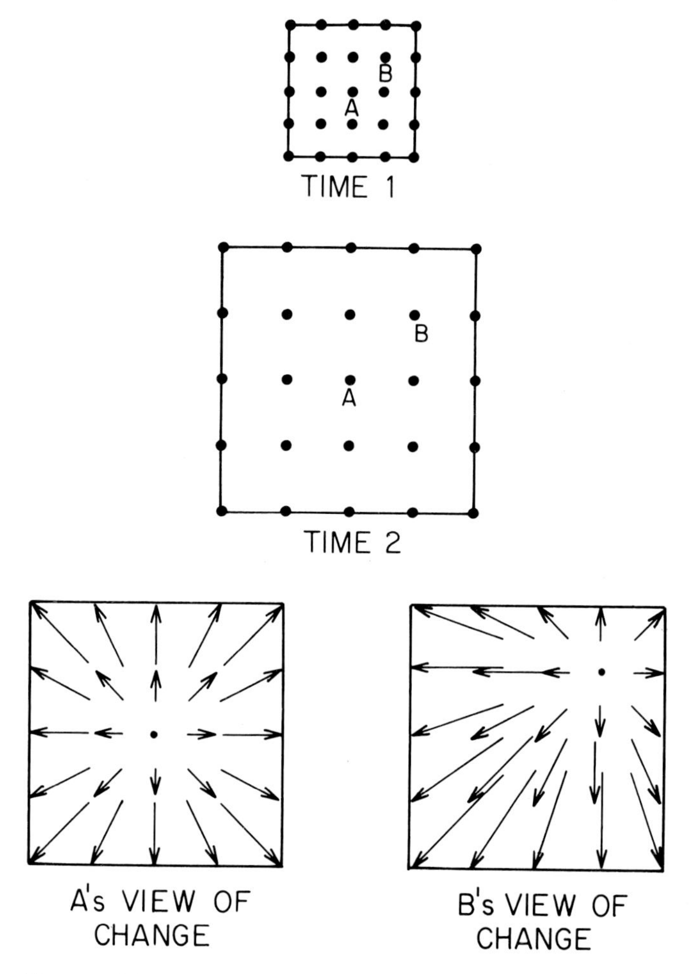

FIGURE 14
Two-dimensional universe expanding isotropically between times 1 and 2. Lengths of arrows represent velocities measured by observers A and B. Each finds (a) all galaxies recede radially and (b) velocities strictly proportional to distance. Both seem at center of expansion.

None can regard himself, therefore, as more centrally located than any other, or necessarily at rest, even though all see every other galaxy in recession. In contrast, a universe that expands according to a different velocity–distance relationship does *not* preserve the same structure and *does* define a unique center.

Even this second example is severely limited: it deals with a flat universe where Euclidean geometry* rules. A common example of a *curved*, two-dimensional, expanding universe is an inflating balloon on the surface of which are uniformly distributed dots. Again, each dot sees every other receding at a velocity proportional to its distance; every dot can be regarded as the center of expansion; and the original homogeneity of the distribution is retained.

The universes described, with velocity proportional to distance, all have the extraordinarily pretty consequence that their essences are eternal. That is, their degrees of uniformity and isotropy do not change, even as their scales enlarge. Furthermore, every observer deduces the same structure and rate of change. Egalitarianism prevails; privilege is nonexistent. Regarding the material universe and the space in which it is arrayed, a perfect democracy exists. Can the biological universe be any different? More to the point, is there but a single location where life exists, the particular one we inhabit?

Expansion today implies a more crowded universe earlier. Working backward in time, a given pair of separating galaxies may once have been in contact. But galaxies having twice the first pair's separation likewise have twice their relative velocity, so they too would have adjoined at the same instant. In fact, all galaxies can be traced in reverse to a common origin. This instant of coincidence can be calculated from Hubble's constant. Just as any two automobiles that have separated at a constant 60 miles per hour to a distance of 180 miles have been traveling for 3 hours, so, too, any two galaxies departing at 10 miles per second (or 300 million—3×10^8—miles per year) to a separation of a million light-years (or 6 million trillion—6×10^{18}—miles) have traveled for 20 billion (20×10^9) years. This time, called the Hubble age, is an upper limit to the actual age of the universe because it is almost certain that galaxies separated more rapidly in the past.

---

*Euclidean geometry is the kind commonly taught in high schools. In it, only a single line can be passed through a point and be parallel to another line. Parallel lines are defined as those that never intersect.

Is there additional evidence for an expanding universe? Indeed, yes. Start from the observation that most of the matter in today's universe is in the form of atomic hydrogen, nature's simplest and lightest element. Our previous retracing of cosmic history has suggested that matter erupted from a common origin at a particular instant, one we have now crudely dated. From everyday experience—for example, hand-pumping a bicycle tire—we know that compression results in heating. The compression involved in (hypothetically) crushing the universe to infinitesimal dimensions would have been so extreme that the accompanying increase in temperature would have been immense. Matter as tightly packed and as fearsomely agitated as projected in this backward glance should, it seems, have fused into elements much heavier than the lightest—*unless*, that is, an intense concentration of energetic radiation was simultaneously present. A furious maelstrom of gamma rays, X rays, visible light, radio waves, and the like, could have torn heavy elements apart as rapidly as they formed. Since such radiation, if once present throughout space, would have no place to escape, it must still be present today. We are thus led from a knowledge of the simple chemical composition of the present universe and the observation of its expansion to the conclusion that a background of radiation must pervade the whole of the universe.

Can we predict the properties of this background? Surely it must come at us (even if we could shift to anyplace at all within the universe) with equal magnitudes from all directions. Its initial omnipresence and its subsequent multiple scattering of matter guarantee a high degree of isotropy. And it must, of course, be cool: expansion, almost twenty billion years' worth, assures that. In fact, a crude temperature prediction of less than ten degrees Kelvin (i.e., ten Celsius degrees above absolute zero) follows from the Hubble age and the knowledge that not all the primordial hydrogen escaped fusion, a quarter or so having merged into helium. Finally, the detailed energy distribution of the remnant radiation can be accurately forecast. In the nascent stages of the universe, radiation and matter interacted so extensively that a detailed balance ensued. On one hand, radiation impinging on some particles knocked them to

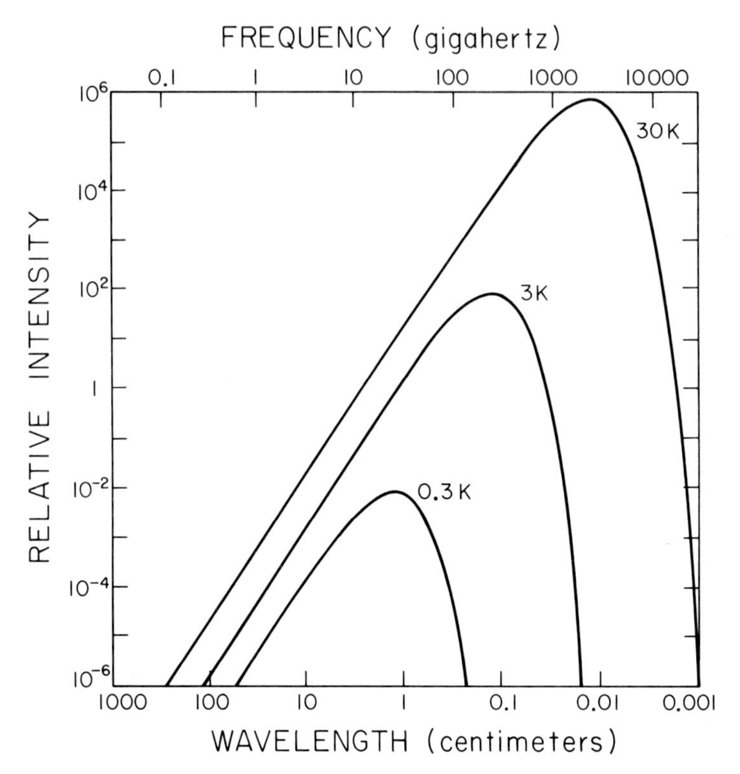

FIGURE 15
Cosmic thermometers. Power radiated by perfect emitters has characteristic dependence on wavelength and frequency. Area under curves and peaks of curves depend upon temperature of emitter. Measurements of power at different wavelengths therefore yield temperature.

new positions, changing their velocities and energies in the process. On the other hand, precisely as many particles were battered from other positions, velocities, and energy states into the ones vacated. Physicists describe this situation as a state of thermal equilibrium, and both experimental and theoretical studies agree on its characteristic spectral signature (fig. 15). The shape of these distributions is the same for all temperatures, but their heights and the wavelengths of their maximum intensities are strongly temperature dependent. Consequently, measurements along curves so shaped provide the astronomer with a cosmic thermometer.

All three predictions—isotropy, cool temperature, equilibrium spectrum—have by now been verified. The first astronomers to sense the archaeologist's thrill at uncovering fossil evidence of our past were Arno Penzias and Robert Wilson, although the significance of their discovery was first recognized by Robert Dicke, James Peebles, and their coworkers (who independently rediscovered the much earlier work of George Gamow and his students). An analogy illustrating how much of the past they sampled is provided by comparing man's observation of the evolving universe with the view from the top of the Empire State Building.* If ground level corresponds to the beginning of expansion of the universe, then the most distant known galaxy is seen down near the fiftieth floor; but the lingering remnant of the universe's fiery beginning is a mere half inch above the street! The painstaking measurement by Penzias and Wilson evidently unveiled a page from the very story of creation.

Their work was done with a highly specialized radio antenna previously used for satellite communications. Its principal virtue was that the entrance field of view was precisely known; unwanted radiation could not sneak in from extraneous directions. By careful calibration and calculation, one could therefore account for all contributions to detected signals: those from cosmic sources, from Earth's atmosphere, from the antenna structure and its immediate environment, from the various electronic components, or from a variety of other sources. But no matter how meticulous their accounting or in which direction their telescope was aimed, some unexplained radiation with the equivalent temperature of a few Kelvin degrees always remained. Moreover, it showed no daily or yearly fluctuation, which suggested an origin beyond the solar system. Nor was any change associated with the overhead passage of the

*The analogy is taken from a 1967 article by P. J. E. Peebles and D. T. Wilkinson, "The Primeval Fireball."

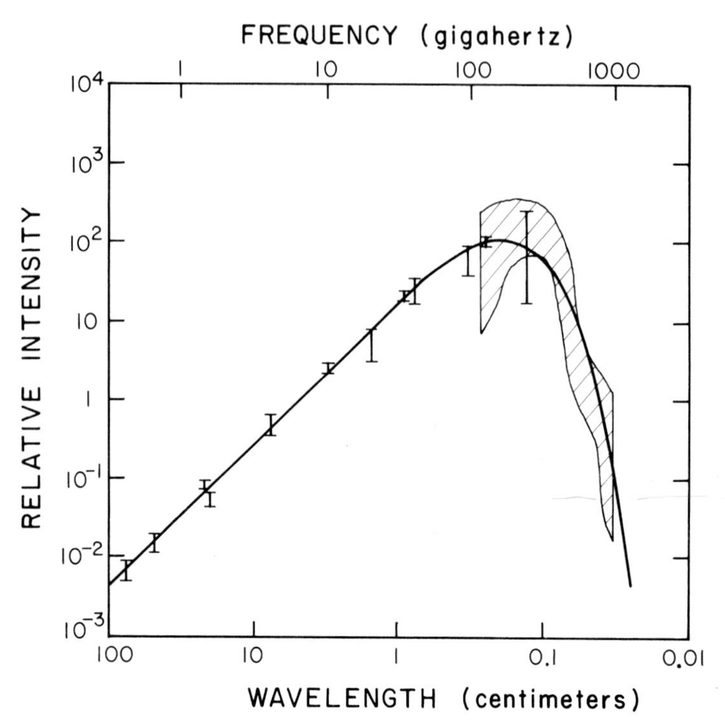

FIGURE 16
Echo of creation. Powers measured (I-shaped lines and hatched area) at various radio wavelengths follow curve expected for perfect radiator near 3 degrees Kelvin (−270° C, −454° F). (Courtesy D. Woody)

Milky Way, thus placing the source outside the galaxy. Planet Earth was apparently bathing in signals arising at great distances. These were quickly recognized to be echoes of creation, remnant whispers of a primordial Big Bang. Perhaps no more significant cosmological discovery has ever been made. The proof of its nature emerged quickly from the series of measurements tracing the signal strength's dependence upon frequency: all agreed closely with the distribution predicted for radiation at a temperature near 3 degrees Kelvin (–270 degrees Celsius, or –454 degrees Fahrenheit).

How should one visualize the Big Bang? As an explosion of space itself rather than as one hurtling matter into a vast preexisting void. While the latter would satisfy Hubble's law, it would not preserve universal homogeneity. Instead,

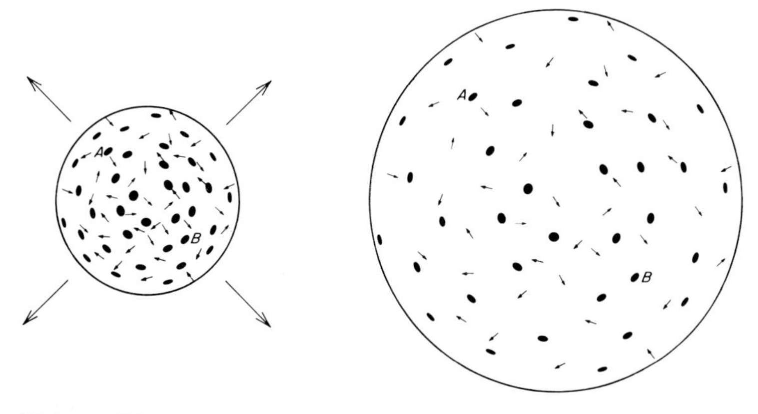

FIGURE 17
Curved, two-dimensional, expanding universe. Dots are galaxies, arrows randomly directed packets of energy (photons). Both A and B reach same conclusions as counterparts in flat space of figure 14. Energy density in photons per unit area declines as universe expands.

the matter sprayed outward would leave a growing hole at the origin, thereby creating a unique cosmic address. Furthermore, any fossil radiation present in today's universe would be segregated from matter, for light emitted during the initial explosion would have fled its origin faster than the more sluggish matter. Since neither is the case, we conclude there was no primordial clump of matter. Nor was there a center to the explosion. Space was, and is, uniformly occupied. There was, and is, no outer boundary to the distribution of matter.

Again, the curved, two-dimensional analogy of an inflating balloon is illustrative. If dots on its surface represent galaxies, we can imagine crawling ants, randomly directed, to be the packets of radiant energy remaining from the universe's fiery birth. Both distributions—matter and energy—are, and will always be, thoroughly intermixed. As the balloon universe expands, all ants, regardless of orientation, find themselves chasing receding dots. They therefore constantly lose energy. This decline in energy carried by each ant, coupled with their growing mutual separations, translates into ever less energy in every given area of the balloon's surface. The declining density of radiant energy corresponds to cooling temperatures. In fact, temperature is a convenient clock for an expanding universe, its rate of decline mirroring the rate of expansion.

Will the cosmic clock tick at the same rate forever? In other words, will the universe continue to expand and cool inexorably at the present rate? Or will it gradually decelerate, slowed by the combined gravitational pull of all its resident matter? If so, is the pull sufficient to reverse the expansion, triggering a future epoch of collapse and reheating? Only by determining these answers can we comprehend the *totality* of time: past, present, and future. The questions are analogous to those surrounding the launch of a projectile from Earth's surface. Was the launch velocity great enough for the projectile to escape Earth's influence entirely? Or would the projectile rise at an ever slower pace, finally stopping completely and retracing, in reverse, its ascent? In between is the possibility of a deceleration so gradual that it brings the projectile to rest only after an infinite time. The critical determinants of which will happen are the mass and size of Earth. Compare, for example, the launch vehicles used in sending astronauts from Earth to the moon with those required for their return. Because of the moon's smaller, lighter bulk, the latter vehicles were relatively puny.

In the same way, the fate of the universe hinges on the mass it encompasses within some given distance—that is, on the mass per unit volume, or mass density. We know the outward speed of a galaxy at every distance, but we can as yet only crudely estimate the mass densities encompassed. Yet is it not wondrous that we even have schemes for, in effect, weighing the universe? How? A favorite (albeit frustrating) method is by extending the Hubble diagram to the greatest distances (hence greatest recessional velocities) within our technical capabilities. Hubble himself described it as stretching for "the dim boundary—the utmost limits of our telescopes. There, we measure shadows, and we search among ghostly errors of measurements for landmarks that are scarcely more substantial."* The reward for such persistence is an earlier view of the universe, to the time when light from the remotest galaxies *began* its earthward journey. If, indeed, the universe is (and was) decelerating,

*Edwin Hubble, *The realm of the nebulae* (New York: Dover Publications, 1958), p. 202.

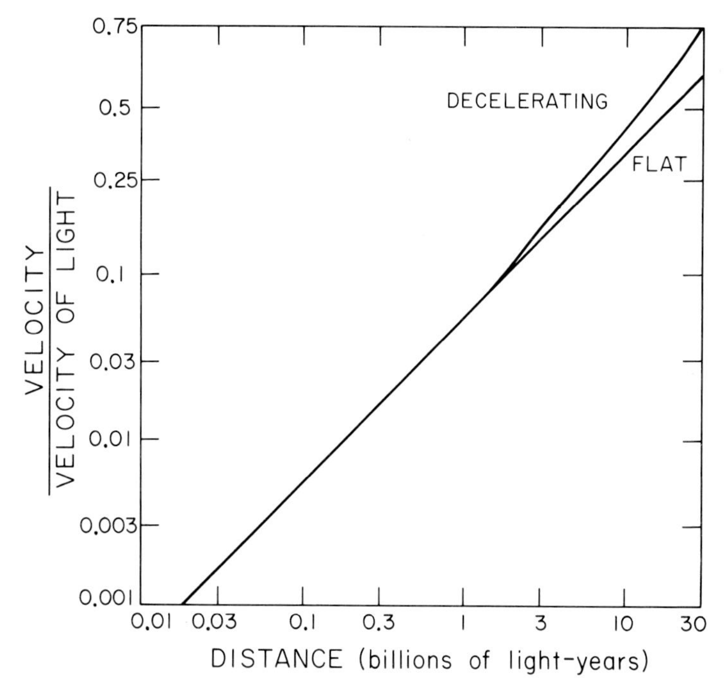

FIGURE 18
Hubble diagram in flat and decelerating universes. At earlier times reached by peering to greater distances, velocities in decelerating universe exceed those in flat universe.

the rate of expansion long ago must have exceeded that today. The ratio of galaxies' recessional speeds to their distances therefore should increase with distance, the Hubble diagram showing not an indefinite straight line but a gradually rising curve. The degree of curvature indicates how drastic the deceleration is or, equivalently, how considerable is the mass density. While the technical difficulties are extraordinary, today's best measurements are consistent with a universe so deficient in mass density that it will expand forever. Such a universe is, for fairly evident reasons, called open and infinite. Its present age is less than the twenty billion years estimated by assuming no deceleration, but not as dramatically less as would be the case in a closed, finite, strongly braked universe.

A second method for weighing at least a portion of the universe is a direct one. First, one chooses a volume of space sufficiently great to be a representa-

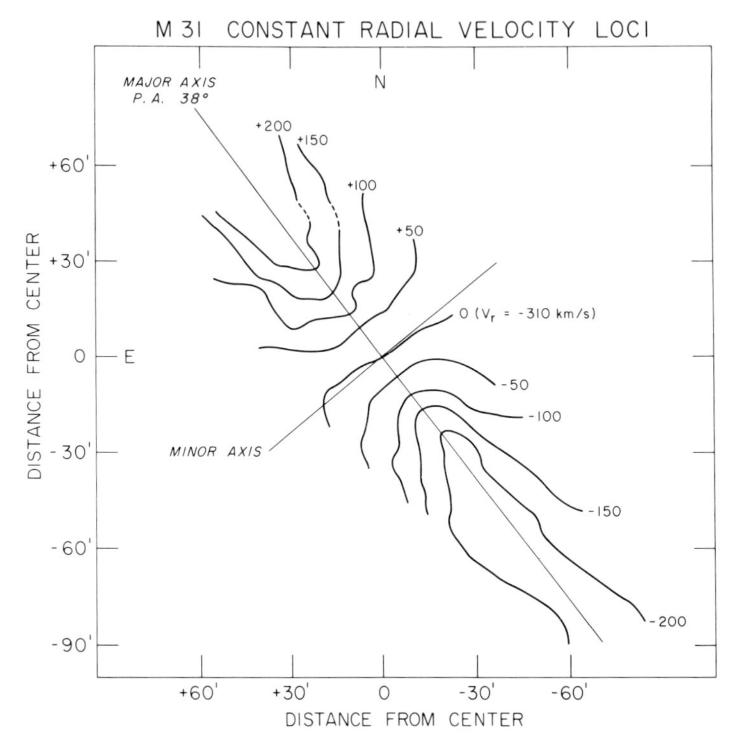

FIGURE 19
Rotation of Andromeda Galaxy (fig. 4). Contours of constant radial velocity—negative approaching, positive receding. High velocities at great distances from galaxy's center imply large mass concentrations. (Reprinted from Sandage, Sandage, and Kristian, eds., *Stars and Stellar Systems*, vol. 9: *Galaxies and the Universe*, by permission of the University of Chicago Press. Copyright 1975 by The University of Chicago.)

tive sample of the universe at large. Then, one weighs the galaxies within and calculates the density by dividing the total mass by the total volume. Weighing galaxies, though a humbling experience, is not particularly difficult. Some orbit one another, their measured separations and orbital periods yielding estimates of their masses through straightforward applications of Newtonian mechanics, just as the sun's mass was long ago determined by analyzing Earth's revolution about it. Other galaxies spin about their own axes. The rotational speeds at various distances from these axes can be measured and interpreted in terms of the mass within the distance spanned. A remote star in a massive galaxy, for example, must circle more rapidly than one equally distant from the center of a puny galaxy to generate the extra centrifugal force needed for balancing the stronger centripetal pull of gravity it experiences. And what do these measurements, when finally converted into densities of matter, tell us? That *luminous* matter, at least, is far too sparse to brake the universe significantly, sparser even than the vacuous four hydrogen atoms per cubic meter which would suffice to close space.*

A third indication of an open, infinite universe comes from searching for trace elements that might have been produced in the early moments following the Big Bang. Some fused products, most notably helium, could have escaped disintegration by the radiative fireball engulfing them. But helium production required merging four atoms of hydrogen, and in a crucial intermediate step two were forged into a unit of heavy hydrogen called deuterium. If the infant universe's matter density had been large, nearly all the deuterium created would have been consumed by helium production; if, however, it had been slight, not all the deuterium would have found partners with which to interact. Some, admittedly a trace, would have escaped subsequent fusion—and would persist to this day. So its present abundance reflects conditions early on; and the best current measurements of deuterium abundance suggest a universe whose density falls well short of that needed to halt expansion.

The universe has apparently happened but once. It is *infinite* in extent, and it will expand *forever*. These at least are the preliminary indications from some exceedingly difficult observations, none of which, by itself, is definitive. If future studies verify these conclusions, the implications are staggering, for who can claim to comprehend infinity and eternity?† We humans, for in-

---

*For comparison, the density of water on Earth is a trillion trillion (10^{24}) times greater.

†One who has tried is cosmologist James Gunn. His 1979 article forms the basis for the speculations that follow.

stance, often believe ourselves unique. But in an infinite setting nothing is unique. Nature has, or will have, ample opportunities to attempt every experiment innumerable times. It follows that somewhere another planet exists identical to Earth where at this instant someone identical to yourself is reading these identical words. In addition, someone else, who differs only in, say, color of clothing, is engaged likewise. In fact, every conceivable variation (and some inconceivable ones as well) *must* exist. Everything supplying majesty and grandeur to our setting—the colors of a rainbow, the innocence of a child, the flight of a hummingbird, the fragrance of a rose, everything—is repeated ad infinitum throughout the universe. Need we any more explicit reminder of our own diminutive finitude?

FOR FURTHER READING

ABBOT, EDWIN A. 1965. *Flatland*. 5th ed., revised. New York: Barnes & Noble, Inc. 108 pp.

BURGER, DIONYS. 1965. *Sphereland*. New York: Thomas Y. Crowell Co. 208 pp.

GARDNER, MARTIN. 1976. *The relativity explosion*. New York: Random House (Vintage Books). Pp. 141–163, 176–193.

GOTT, J. RICHARD III; GUNN, JAMES E.; SCRAMM, DAVID N.; and TINSLEY, BEATRICE M. 1976 March. Will the universe expand forever? *Scientific American* 234(3):62–79.

GUNN, JAMES E. 1979. Observations in cosmology: the shape of space and the totality of time. *The great ideas today. 1979*. Mortimer J. Adler, ed. Chicago: Encyclopedia Britannica, Inc. Pp. 8–32.

HENRY, PAUL S. 1980. A simple description of the 3K cosmic microwave background. *Science* 207:939–942.

HUBBLE, EDWIN. 1958. *The realm of the nebulae*. New York: Dover Publications Inc. 207 pp.

MULLER, RICHARD A. 1978 May. The cosmic background radiation and the new aether drift. *Scientific American* 238(5):64–74.

PEEBLES, P. JAMES E., and WILKINSON, DAVID T. 1967 June. The primeval fireball. *Scientific American* 216(6):28–37.

SILK, JOSEPH. 1980. *The big bang*. San Francisco: W. H. Freeman and Co. Pp. 61–126.

4
ATOMIC EVOLUTION

I am, in point of fact, a particularly haughty and exclusive person of pre-Adamite ancestral descent. . . . I can trace my ancestry back to a protoplasmal primordial atomic globule.

—W. S. GILBERT, *The Mikado*

Archaeologists deserve immense respect for their skillful reconstructions of ancient civilizations from the merest shreds of information: a few teeth, a fragment of bone, some imperfectly preserved footprints, scattered flakes of stone. But their courageous extrapolations pale to insignificance beside the panache of astrophysicists, who convert fragmentary knowledge of current cosmic conditions into detailed scenarios of conditions in the first moments after the Big Bang, moments that are separated from us by a temporal chasm approaching twenty billion years! Are such efforts foolhardy in the extreme? Or is our knowledge of laws of physics sufficiently precise, and our faith in their eternal veracity strongly enough warranted, to justify the attempt? Only future observation will tell, but to forego the effort would betray a depressing absence of curiosity. We are, after all, the first generation equipped with enough knowledge to remove this ultimate historical reconstruction from the realm of dreamy speculation.

Whatever the details of the changes during the first few moments of the universe, we can be sure that they occurred at a faster pace then than now. Recall that the expansion of the universe was offered in the previous chapter as the variable quantity against which the passage of universal time could be judged; and the rate of that expansion depended on the overall outward momentum of the total system—that is, on the balance between the outward violence of the initial explosion and the density of the matter tugging to restrain it. Very early in the universe's history, when gravitational braking had as yet had little effect, events followed one another rapidly; later, as gravity's restraining influence took hold, the pace slowed. As a result, time intervals of equal duration have not bracketed changes of equal magnitude. Time itself fails as a uniform measure of progression. We must resort to a more appropriate parameter, temperature.

Temperature has the attractive feature that it scales inversely with the size of the universe: shrink the universe to, say, one-tenth its present size, and the prevalent temperature increases by the same factor of ten. But we are not

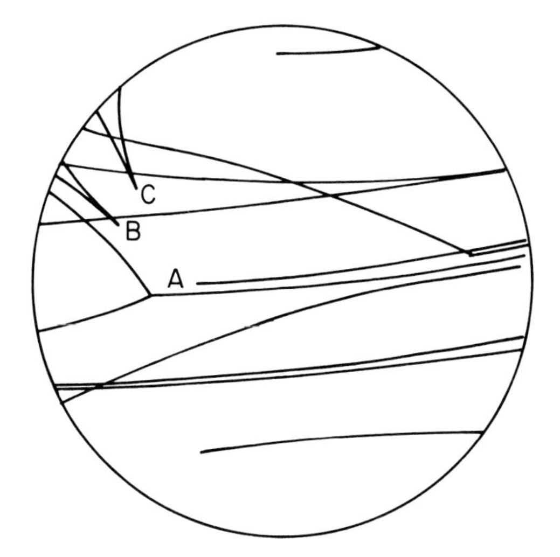

FIGURE 20
Particle creation and annihilation. At point A, particle from accelerator collides with proton. Among fragments are two trackless packets of energy that, at points B and C, create pairs of positive and negative electrons. (Adapted from figure 3.2 of Kenneth W. Ford, *The World of Elementary Particles* [New York: Blaisdell Publishing Company, 1963].)

toying with mere factors of ten, for we wish to trace our ancestry to its very beginning roughly twenty billion years in the past. To do so we must multiply the temperature of the radiation lingering in today's universe by tens of billions to estimate conditions in the first year or so of cosmic history: today's three Kelvin degrees was once hundreds of billions of degrees—even more!

Such numbers bewilder the uninitiated. They exceed by ten thousand times even the temperature at the center of the sun, where we know nuclear reactions of extraordinary ferocity take place. But the numbers are within the working experience of those nuclear physicists who operate particle accelerators to study the elementary constituents of matter. They tell us, with confidence, that the distinction between matter (real substantive chunks of "stuff") and radiation (pure packets of energy in the form of oscillatory electromagnetic signals) becomes blurred at these enormously inflated temperatures: from pure energy, pairs of particles can spontaneously materialize; likewise, existing pairs can annihilate one another, leaving in their wake brilliant flashes of X rays or gamma rays (extremely short-wavelength forms of light). Matter and energy are, it seems, merely alternative manifestations of the same entity.

The physicists, however, are not omniscient. Their ignorance of the properties of some exotic particles, those that appear only at temperatures of trillions of degrees, and of the details of the interactions among them, obscures their vision with regard to times when the universe was mere thousandths of a second old. Furthermore, uncertainties intrinsic to the subatomic world will probably blind us forever to events at *the* beginning.

Events in the immediate posttheological era, however—after the command "Let there be light" had been obeyed—are decipherable. Elementary particles, the fundamental constituents of matter, were created and destroyed simultaneously with their mirror images, called antiparticles. Soon they equaled in number the photons, the packets of energy in which light manifests its presence. A perfect equilibrium between particles and photons was established. But the temperature was falling rapidly as the universe expanded at a rate faster than it ever again would. Below certain thresholds, some particle-antiparticle pairs could no longer be created: the energy latent in the cooling radiation field had fallen below the minimum necessary to account for their masses even when stationary.* As time's passage brought dwindling temperatures, particles were "frozen out" of the energy-matter interchange in reverse order to their masses.

The constituents of matter most critical in synthesizing the succession of heavier chemical elements, hence eventually life, are the neutrons and protons. These near twins—one neutral, the other electrically charged—combine in various numbers and ratios to define the nuclei of all atoms. They are sufficiently massive that their threshold temperatures were descended through at so early a cosmic age that ignorance still shrouds our perception.

*Every particle has a so-called rest mass, m_o, endowing it with an equivalent energy content, E, given by Einstein's famous relation, $E = m_o c^2$, where c is the speed of light.

TABLE 3 **Particles of the Early Universe**

Particle	*Particle mass* / *proton mass*	*Particle charge* / *proton charge*	*Threshold temp. (billions Kelvin degrees)*	*Mean life (seconds)*
Photon	0	0	0	stable
Electron	0.0005	−1	5.93	stable
Proton	1.0000	+1	10,888	stable
Neutron	1.0014	0	10,903	920
Deuteron	1.9990	+1	21,765	stable
Helium nucleus	3.9724	+2	43,252	stable

All we know with any confidence is that, by the time the universe was a hundredth of a second old, equal numbers of neutrons and protons coexisted with approximately a billion times more photons. Although the two types of nucleons strongly attract each other and themselves, they do so only when brought into near contact. Fusion into heavier nuclei therefore requires extremes of both density (for purposes of crowding) and temperature (so that the particles may be flung at each other with adequate vigor). The appropriate conditions persisted only during a "temporal window" whose age and duration could each be measured in minutes. At the beginning of the window, the determining factor was the temperature of the photon bath; if too hot (hotter than about ten billion degrees Kelvin), the photons were energetic enough to break up bound groups of particles. Posing the end of the window was the spontaneous decay of free neutrons (those not bound into nuclei) into protons, about half of a given collection undergoing this conversion every fifteen or so minutes. Consequently, if fusion of protons and neutrons had been delayed too long, it might just as well have waited forever, since no neutrons would have remained.

Neutron-proton fusion into deuterons, nuclei of heavy hydrogen, began when the universe was between three and four minutes old. By that time only thirteen neutrons remained for every eighty-seven protons. It follows that, for a long time thereafter, seventy-four of every hundred nuclei remained pure hydrogen, while only twenty-six became nuclei of heavier elements. How

TABLE 4 **Helium Abundances**

Location	*Percentage helium by mass*
Solar neighborhood	26–32
Planetary nebula (one object)	42
Other normal galaxies	22–34
Dwarf galaxies (two objects)	29

heavy? Well, if conditions were suitable for the fusion of deuterons, they were even more amenable to subsequent fusions ending at the extremely stable helium nucleus, which consists of two protons and two neutrons. That is, having surmounted the barrier of forging a fragile two-nucleon nucleus, the steps to a four-nucleon variety were easy. But beyond that, they were insurmountable: no stable nucleus exists with five nucleons; as quickly as one forms, it disintegrates. Chemical production ended accordingly after less than thirty-five minutes of cosmic history. By then, universal expansion had so rarified and cooled the substrate that nuclear fusion ceased (temporarily, as we shall see in the next chapter).

Have we any check that these immense temporal extrapolations accurately reflect reality? Indeed we do—all around us. Ubiquitous helium exists in an abundance that varies little from place to place. The same 20 to 30 percent (by weight) is found wherever we look. This accords, within experimental uncertainty, with the quantity predicted by Big Bang cosmologists. In contrast, the distribution of scarcer, heavier elements varies dramatically. A plausible explanation is that helium production predated the formation of stars and galaxies whereas production of heavier elements did not. In support of this hypothesis, calculations of the total energy released from the conversion of hydrogen into the presently known quantity of helium reveal an excess by at least ten times over the energy radiated by all galaxies since their formation. If this is so, then helium synthesis could not have occurred only within galaxies; it must have preceded their formation.

So far we have concentrated only on neutrons and protons. This is proper, since these ingredients determine the ultimate breed of an atomic

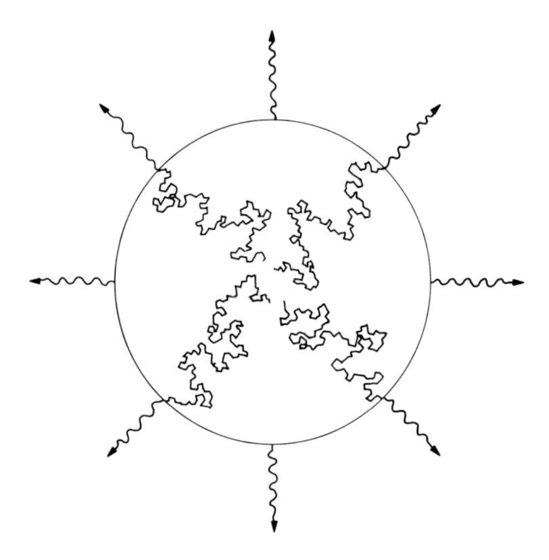

FIGURE 21
Energy escapes the sun. Photons produced near sun's center suffer repeated collisions with electrons during fifty-million-year journey to surface. Resultant radiation pressure prevents collapse.

nucleus. But matter would be unrecognizably different if the only electrically charged particles within its constitution were positive. The universe, however, was also endowed with negative charges called electrons, fragments of matter whose mass is a mere two-thousandth of a nucleon's. Their threshold energies and temperatures are correspondingly reduced, so that pair creation of electrons and antielectrons (called positrons) can persist long after creation of protons and neutrons has ceased. Even so, the productive epoch lasted scarcely ten seconds, after which the mirror pairs could only annihilate. Had the numbers of electrons and positrons been exactly equal initially, none would exist today. But in fact electrons very slightly outnumbered positrons—by perhaps one in a billion. This minute fraction endured from age thirty-five minutes onward, joining a precisely equal number of protons to make the overall electrical charge of the universe zero.

Photons interact vigorously with free electrons, but much less so with those that are bound into orbits surrounding atomic nuclei. Views of the sun (fig. 8) and of the depths of space (fig. 3) provide examples of the two types of photon-electron interaction. In the sun, photons produced at the center struggle for fifty million years to escape from its surface. Multiple scatterings by the sea of free electrons enveloping the solar core effectively buffet and trap the emergent flux for such periods. As a result, we see only the outermost layer of the sun, the surface of last scattering. The sun's innards are impenetrable. Contrast this opaqueness with the crystalline transparency of deep space. Photons reach us, evidently undeflected, from galaxies ten billion light-years distant. Jostling by free electrons never happens. But that is today. What of yesteryear?

Although attractive forces lure electrons toward protons, actual capture into bound orbits is extremely unlikely if their encounters occur at excessive energies. Man-launched planetary probes offer an analogy. The Voyager spacecrafts both spun past Jupiter on ricochet courses to Saturn. The planned Galileo spacecraft, however, will be trapped by the giant planet, adding an artificial satellite to its sizable retinue of natural ones. The difference in fates is due primarily to the energy with which the crafts approach their potential captor. In the case of electrons and protons, the energy of interaction is determined by the photon bath in which the particles are immersed. Only when the bath has cooled to about three thousand Kelvin degrees are electrons lethargic enough to combine permanently with protons to make neutral atoms. At this epoch of combination, the universe switches from opaque to transparent. The age of the universe at which this critical temperature is reached is seven hundred thousand years.* Matter and radiation are forever after decoupled and subsequently follow independent courses. Matter is at last freed to condense into galaxies and stars. Yet another twelve or eighteen billion years must elapse before creatures who can reconstruct these milestone happenings will appear.

*Note the irony of using years, a unit of time representing the completion of one full revolution of Earth about the sun, long before either exists.

Negative electrons combining with bare positive nuclei to form neutral atoms signaled the onset of a new era in the evolution of matter. Recall that photons were a billion times more numerous than particulate matter. Therefore, when the two ceased interacting, the pressure exerted upon matter dropped by the same colossal factor. Even so, accretion of the particles into massive concentrations required an additional happenstance, the appearance of instabilities. If all the matter in the universe had remained perfectly uniformly distributed forever, no condensation could have occurred. But if, instead, random fluctuations had produced some exceptionally dense regions, these might have become gravitational focuses upon which self-collapse could have been centered.

Whether or not collapse ensued depended upon the relative magnitudes of two oppositely directed forces: the inward pull of gravity versus the outward thrust of internal pressure. The magnitude of the gravitational effect is larger for big concentrations than for small ones, whereas the opposite is true for dispersion by pressure. Physicists can quantitatively assess the outcome of the inward-outward tug-of-war between gravity and pressure. How large must a density fluctuation be in order that its weight can overwhelm its internal pressure? The answer is that the mass of at least a million suns must be present—and, indeed, the masses of the smallest galaxies we observe, the so-called dwarfs, are of this magnitude. This minimum galactic mass equals the combined contributions of a thousand trillion trillion trillion trillion trillion (10^{63}) hydrogen atoms!

As each protogalaxy collapsed, it also rotated about some axis; the tidal tugs of neighboring protogalaxies assured that. Just as an ice skater spins faster when tucking her arms closer to her body, so too does a shrinking galaxy spin ever more rapidly. Matter falling along the rotational axis encounters little resistance, but that collapsing perpendicular to it must overcome the centrifugal force common to all spinning objects. The result is a high proba-

FIGURE 22
Dwarf galaxy in Sextans. (Palomar Observatory photo)

bility of flattened, or pancake-shaped, systems, the amount of flattening depending upon the intrinsic quantity of rotation. Hence various structural types emerged, populating the cosmos with objects of breathtaking beauty. The most rapidly rotating systems were so effective at keeping matter centrifugally scattered that the formation of dense subconcentrations within them required long time spans. In a sense, they were (and are) conservative in the consumption of their initial material endowments. Other, slower rotators were more spendthrift; early in their histories stars and clusters of stars formed within them, leaving little debris from which subsequent generations of stars could coagulate. We therefore observe galaxies of different types hosting stellar families of predominantly different ages and containing varying amounts of residual interstellar matter.

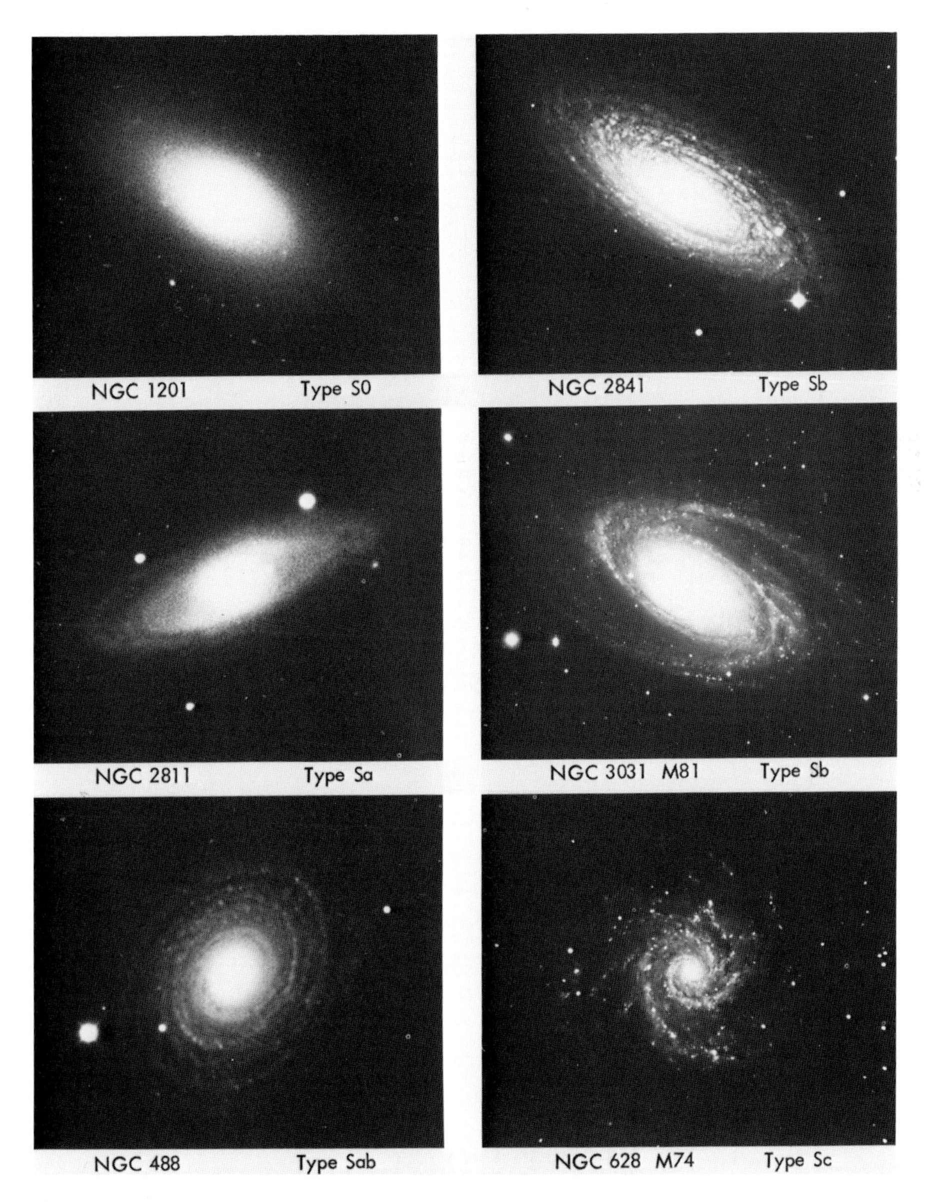

FIGURE 23
Rotating, spiral galaxies. Types Sc are richest, and S0 poorest, in young stars and interstellar matter. (Mount Wilson and Las Campanas Observatories, Carnegie Institution of Washington)

Although we have traced a mere seven hundred thousand years of cosmic history, already it is apparent that our existence is a fortuitous circumstance. Or is it? Certainly one can easily imagine universes where no creatures ever appear who can contemplate their ultimate origins. In fact, said universes are so easily imaginable that perhaps the case can be restated: the fact that we are here means the universe has to be the way it is. What changes might have prevented our appearance?

To begin, consider the slight original excess of matter over antimatter. No one knows why one type should outnumber the other; both are equally satisfactory as foundations for larger structural units. In fact, our sense of symmetry is offended by the numeric inequality. In a truly elegant and aesthetic universe, neither would be favored over its mirror image. But had such been the case, all matter and antimatter would have mutually annihilated in those early, crowded moments of cosmic history, leaving only radiation. No galaxies, stars, planets, butterflies, armadillos, or people would ever have appeared.

Likewise, the precise equality between positive and negative electrical charges seems necessary for our existence. The dominant force at work on the large scale in our present universe, organizing it into the shapes and forms we see, is gravity. But this force is exceedingly weak compared with that of electromagnetism. If, for example, the sun and Earth differed from electrical neutrality by a single part in a trillion trillion trillion (1 in 10^{36}), then the resulting mutual electrical attraction or repulsion would overwhelm the gravitational attraction that binds one in orbit about the other.

What if the initial rate of expansion of the universe had been different? Consider as one alternative an especially violent, explosive origin hurtling matter rapidly outward, combined with a reduced initial density of matter. Under these conditions the matter is dispersed too fast for much of anything to happen: no element synthesis; no gravitational pockets collecting the stuff to form stars and galaxies; just elementary particles scattered hither and yon, lonely in their expanding isolation. By contrast, a universe that unfolded too slowly and had too dense an initial concentration would suffer the opposite fate: all the protons and neutrons present at the start would be cooked into heavier elements immediately, before any neutrons had a chance to decay into protons. As a result, the sources of energy that power living systems today, as the sun energizes life on Earth, would have been completely exhausted. In some way, by chance or otherwise, our universe managed to thread the narrow line between these two extremes, a circumstance without which we would never have been.

We are also consequences of the original universal isotropy and homogeneity. Had the universe been grossly inhomogeneous, the dense parts would already have burned all their fuel into the ash of heavier elements, eliminating subsequent energy sources. In the rarefied regions, gravitational concentrations would never have acquired sufficient mass to have collapsed into normal astronomical objects. A highly anisotropic expansion would have had the same preventive impact: shearing forces would have been so pervasive in assemblages of protogalactic dimensions that collapse upon some center of symmetry would have been out of the question.

Perhaps economists among the readers wonder about the cost effectiveness of the universe. Need it be so big and incorporate so much mass? Or could a smaller investment have yielded a bigger percentage return? The answer is that our universe does not appear extravagant. Indeed, a smaller investment might have yielded *no* return. If, instead of having the mass for a hundred billion (10^{11}) stars in a hundred billion (10^{11}) galaxies, there had been only enough for a hundred billion stars in, say, one galaxy, the entire evolutionary cycle of that system would have been measured in years, not billions of years. Consequently, there would have been insufficient time to form stars, let alone to ignite the nuclear furnaces within them. Because no heavy elements could have been formed, no living systems (at least none that we can imagine) could have evolved. Apparently the universe must be big and massive if we are to be part of it.

Either we are extremely improbable accidents, the product of conditions and events that could just as easily never have happened, or conditions began and subsequently evolved in just such a way that eventually something would emerge which could reconstruct the entire story. Science will probably never answer whether indeed there could be a universe if there were never anything to contemplate its existence. But for whatever reason—be it chance or preordination—the fact that we can contemplate the totality of space and of time carries with it the obligation that we must do so.

FOR FURTHER READING

DAVIES, PAUL. 1981 May/June. The anthropic principle and the early universe. *Mercury* 10(3):66–77.

FIELD, GEORGE B.; VERSCHUUR, GERRIT L.; and PONNAMPERUMA, CYRIL. 1978. *Cosmic evolution*. Boston: Houghton Mifflin Company. Pp. 243–263.

GALE, GEORGE. 1981 Dec. The anthropic principle. *Scientific American* 245(6):154–171.

PENZIAS, ARNO A. 1979. The origin of the elements. *Science* 205:549–554.

SCIAMA, D. W. 1971. *Modern cosmology*. Cambridge: Cambridge University Press. Pp. 149–203.

SILK, JOSEPH. 1980. *The big bang*. San Francisco: W. H. Freeman and Co. Pp. 127–180.

TRIMBLE, VIRGINIA. 1977 Jan./Feb. Cosmology: man's place in the universe. *American Scientist* 65:76–86.

WEINBERG, STEVEN. 1977. *The first three minutes*. New York: Basic Books, Inc. 188 pp.

WHEELER, JOHN ARCHIBALD. 1974 Nov./Dec. The universe as a home for man. *American Scientist* 62:683–691.

5
CHEMICAL EVOLUTION

When we try to pick out anything by itself, we find it hitched to everything else in the universe.

—John Muir

The initial billion or so years of cosmic history spawned formation of the first structural units of matter, the galaxies and the clusters in which they reside. The chemical inventory bequeathed, however, was decidedly meager: mainly hydrogen with a sprinkling of helium. We who would dare to recreate our past can, with hindsight, see both advantages and disadvantages to the early absence of chemical diversity. A disadvantage, on the one hand, is that no living system, at least none we can think of (a recurring qualifier that may be a result as much of our ignorance as of nature's inability), can be manufactured from such simple ingredients. Helium in particular is notoriously asocial, its internal structure being so symmetrical and self-fulfilled that it shows little affinity for external bonding. Consequently, neither it nor hydrogen can constitute the backbone upon which atoms can arrange themselves into the complex molecules of which life seems to be composed. On the other hand, an advantage of postponing element synthesis is that ample fuel is retained in reserve, fuel that, when eventually consumed, can provide the energy necessary to sustain life.

It is clear, then, that if creatures are ever to arise who can reconstruct their origins, another period of element formation must take place—indeed, must still be taking place. The likely mode of constructing heavy elements will again be to fuse them from lighter ones. To do so will require higher temperatures and densities than can be achieved over volumes the dimensions of galaxies; some coalescence into tighter units must take place within them. The units that emerge to satisfy these demands are the stars.

Stars cycle through rhythms of birth, maturation, and death just as living systems do. But, of course, living systems reproduce their likenesses, an action unlikely for a physical system as nonbiological as a star. Or is it so unlikely? Often the violent death of a star generates compressive forces that sow the seeds of gravitational instability on which new stars can be built. As with life, the cycle apparently closes upon itself, death being simultaneously an ending and a beginning.

Of course, no one will ever witness any single star's complete cycle: our lives are a much smaller fraction of a star's life than that of a mayfly is of ours (mayflies linger at most one full day). By the same token, none of us studies the life cycle of a tree by watching a single specimen proceed from seed to deadfall. Instead, we visit a forest, noticing among the multitude those of all sizes, hence ages. Likewise, by viewing many stars we can catch some at almost every stage of development.

Those stages that are scarcest are the formative ones. We deduce, therefore, that these must occupy exceedingly small fractions of any star's lifetime. Similarly rare are dead stars, whose only source of light derives from the heat retained by the corpse after its internal furnace has expired. In this case, it is the diminished luminosity that makes such remnants difficult to detect. Only every few centuries (although it would be more often if our galaxy were more transparent) do we catch a massive star at its moment of expiration, an explosive event that may illumine the sky for some months. Obviously this too is a brief interval in a star's existence.

Then what stellar stage is most commonly observed, consequently most enduring? It is the steady-state stage where the sun currently resides, in which the energy bubbling outward from its center exactly counterbalances the oppressive weight eternally poised to fall inward. Intuition suggests, and observation confirms, that the greater the weight needing support—that is, the more massive the star—the more furious must be the central release of energy. As a result, the heaviest stars burn brightest. In fact, their consumption of fuel is so rapacious that their initial supply is more rapidly depleted than is the case in smaller stars, even though the smaller ones have less fuel available. It follows that the very brightest stars observed (brightest in the intrinsic sense, not as a result of nearness) are also the youngest, their total lifetimes being so short that to be still radiating they must have formed only a short time ago.

The sun "burns" fuel at an average rate: every second, 600 million (600×10^6) tons of hydrogen are converted into 596 million (596×10^6)

TABLE 5 Stellar Properties

$\frac{\textit{Mass}}{\text{sun's mass}}$	$\frac{1}{10}$	1	10
$\frac{\textit{Luminosity}}{\text{sun's luminosity}}$	$\frac{1}{500}$	1	2,500
$\frac{\textit{Radius}}{\text{sun's radius}}$	$\frac{1}{8}$	1	5
Effective temp. (Kelvin degrees)	1,800	5,800	18,000
Approx. lifetime (billions of years)	4,000	12	0.04

tons of helium, the difference appearing in the form of radiant energy. Although this appetite is truly prodigious, the hydrogen stockpile from which the sun draws is formidably bountiful. There is easily a quantity sufficient to add another five billion (5×10^9) years of undiminished brilliance to the equal span that has preceded the present. Life on Earth is indeed fortunate that the sun had the size it did, and thus the metabolism it had, for advanced life was slow in unfolding. Had the sun been granted a greater hydrogen endowment, it would have spent its fury before life had had much opportunity to develop.

When reconstructing the first moments of the universe, we encountered the fusion that powers the stars. As two protons approach each other, they are mutually repulsed by the positive charge each carries. The situation is analogous to cars on opposite sides of a deep canyon approaching steep hillsides before the brink. If their speeds are insufficient, the cars will not surmount the intervening barriers; if, however, they have sufficient speeds, they will ascend the hills and plunge over cliffs to be trapped in the canyon below. Only a considerable expenditure of energy can then break up the entrapped pair of vehicles.

In the case of nuclei, the speed necessary to overcome the electrostatic barriers before plunging into the viselike grip of the internuclear force requires high temperatures. In addition, the prospect of a pair approaching closely demands high densities. Both requirements are satisfied if an immense

FIGURE 24
Globular cluster of stars. Gravity at work. (Palomar Observatory photo)

collection of atoms can fall together upon a common center. How immense? The planet Jupiter, at a thousandth the mass of the sun, is too small—but not by much. Another factor of ten to twenty in mass would have provided it with the crushing overburden with which to energize its central nuclei to fusion-sustainable temperatures and densities. At the other extreme, where masses exceed the sun's by factors of fifty to a hundred, the collections become too bulky to remain single homogeneous units. They have instead a few, or several, regions of exceptional density, each capable of centering its own gravitational trap. The result is fragmentation into a cluster of stars, not the formation of a single giant anchoring all the matter.

Atoms within stars arrange themselves as do stars within globular clusters: density peaks at the very center and tails off rapidly in all radial directions. Different physical processes therefore take place at different depths within a star. The center, or core, is the furnace, the site where quartets of hydrogen atoms are fused into lighter and more tightly bound units of helium. The resulting material decrement is transformed into an equivalent quantity of energy. In the process of fighting its way from center to surface, this energy pushes against the matter it encounters, effectively preventing its inward descent (see fig. 21).

If the furnace is extinguished—as happens when, for instance, all the central hydrogen is consumed—the descent of peripheral matter can resume, raising the core's temperatures and densities to ever more ferocious levels.

TABLE 6 **Fusion Reactions in Sun's Core**

Repetitions	*In*	*Cycled*	*Out*
2	proton+proton =	deuteron +	scruff +energy
2	proton+	deuteron = light helium	+energy
1		light helium + light helium =	common helium +2 protons +energy
Gross	6 protons		common helium +2 protons +energy
Net	4 protons		common helium +energy

scruff = antielectron (positron) + massless, neutral particle (neutrino)

light helium = nucleus containing 2 protons and 1 neutron

common helium = nucleus containing 2 protons and 2 neutrons

Note that the residual particles—in this case, helium nuclei—possess both greater mass and greater electrical charge than the matter from which they were created. They will consequently be more sluggish against efforts to accelerate their speeds; yet they must overcome higher barriers of electrical repulsion before feeling any mutual attraction. Much greater temperatures must therefore be attained before fusion reactions can be ignited among them. Whereas ten million (10^7) Kelvin degrees (or eighteen million Fahrenheit degrees) sufficed for the fusion of hydrogen into helium, fusion of the latter into carbon requires two hundred million (2×10^8) Kelvin degrees ($3.6 \times 10^{8\circ}$ F). Ignition temperatures ascend rapidly every synthesizing step thereafter. Less massive stars are consequently unable to complete as many successive stages of element synthesis as are their heavier counterparts, simply because their total gravitational potentials cannot generate adequate internal heat.

A star is evidently a collection of atoms freely falling on one another until their increasingly violent interactions release energy adequate to brake (at least temporarily) further collapse. But what triggers protostar formation? As in the case of galaxy formation, inhomogeneities are involved. A collapsing gaseous cloud of galactic dimensions can scarcely be entirely uniform. All atoms will not, therefore, feel pulled toward a single center. Instead, each will realize a prevailing tug toward a neighboring region of above-average density. Each atom that succumbs to the tugs from a particular region enhances that region's attraction for still more atoms. Eventually the aggregation may reach stellar proportions. Since this may occur early in the history of any galaxy as well as any time thereafter, stars of all ages exist simultaneously.

To see where star formation is actively taking place in our own galaxy, one should look at the vast gaseous clouds that populate the interstellar medium. There the echo rebounding from the explosive death of a nearby star may trigger compressive forces squeezing new stars into existence. Short-wavelength energy pouring off young stars excites the gas throughout the

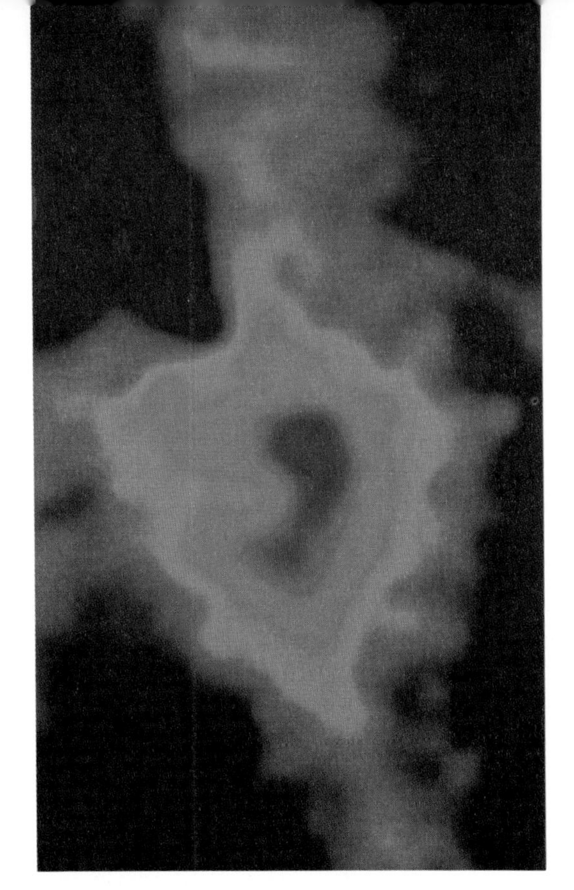

FIGURE 25
"Radio photograph" of carbon monoxide emission (wavelength 2.6 millimeters) from cloud just behind Orion Nebula. Red denotes most intense emission; blue, least intense. Molecular cloud surrounds region of very recent, or even current, star formation. (Courtesy N. Z. Scoville, F. P. Schloerb, and P. F. Goldsmith, University of Massachusetts at Amherst)

cloud that spawned their births, causing it to glow somewhat in the manner of a neon sign. A close inspection often reveals dark patches where the density is so great as to be opaque to visible radiation. These are stellar wombs, sites where protostars pass their embryonic stages. That the pulse of galactic life, as signaled by the birth and death of stars, continues to beat in the Milky Way Galaxy should be a source of rejoicing for us and any other inhabitants within: for surely the freshness of continuous renewal is preferable to the decadence of silent stagnation.

The final stages of a star's existence are among its most fascinating. The onset of death begins when fusion of hydrogen into helium ceases in the core. As mentioned above, this forces an internal readjustment, presaging a succession of fusion processes that use the ash from the cycle just ended as the fuel of its successor. After several cycles, each completed more rapidly than its predecessor, the star acquires a stratified structure, with deeper layers consisting of heavier elements. A very massive star might, for example, have a central core of silicon that is ringed by a shell of oxygen, neon, and magnesium, above which lies a shell of carbon and oxygen, around which is wrapped helium, all within an outer envelope of hydrogen. In lighter stars, the synthesis of elements may cease at an earlier stage. Whatever the mass, however, there comes a time when thermonuclear fusion ceases altogether. For the star, the result is catastrophic. The core implodes, violently severing its connection with the enveloping layers. The latter are wafted off—almost literally evaporated—into the space from which the star originally condensed. With them goes a chemistry vastly enriched over that present at the star's inception. Each generation of stars therefore bequeaths a more chemically diverse environment to its offspring. We would not be, had not several stellar generations preceded the solar system's formation.

Of course, each generation irrevocably removes some matter—that locked in the remnant cores—from the pool of potential stellar material. The cores slowly decline toward a final state as clinkers, the properties of which range from merely bizarre to totally fantastic. In stars roughly the size of the sun, the corpse shrinks to an object with planetary dimensions. This compaction of solar masses into earthlike frameworks results in extravagant densities. A single teaspoon could scoop a ton or so of matter, the approximate weight of a small automobile, from the carcass of one of these former luminaries.

The remains of stars once as glorious as the sun are called *white dwarfs*: "dwarfs" because of their comparatively diminutive stature and "white" because of their lingering high surface temperatures. Just as the shutdown of a

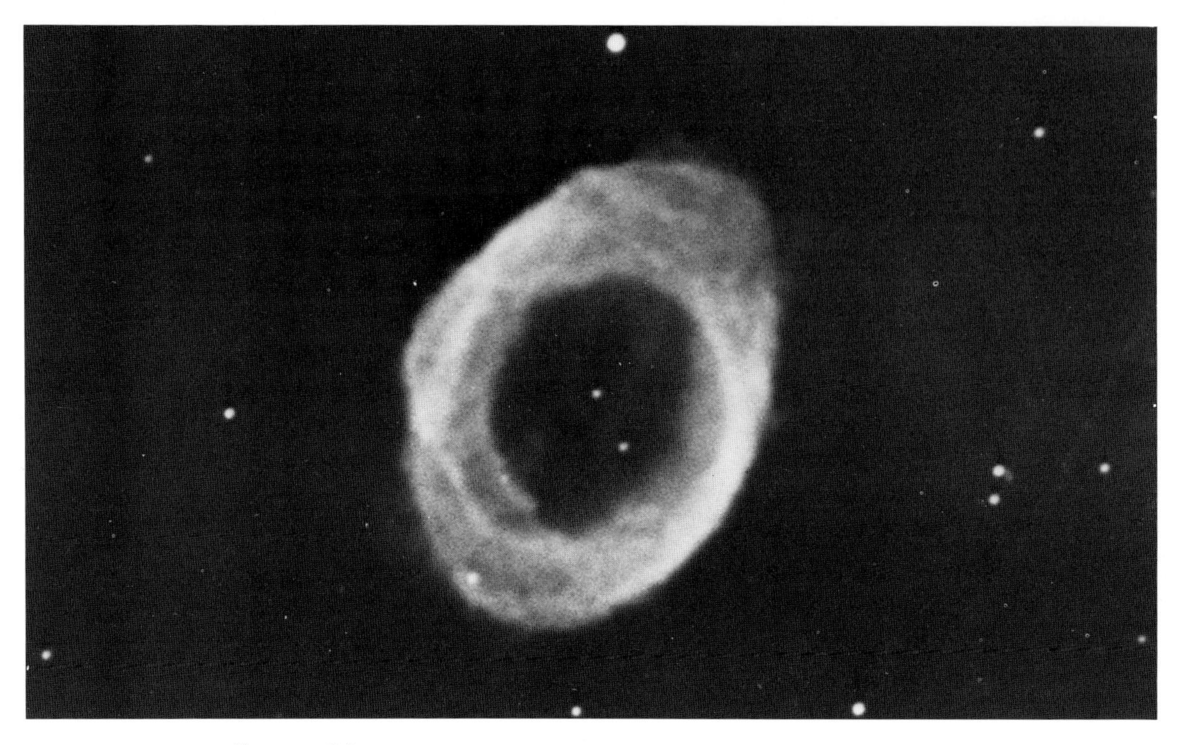

FIGURE 26
Ring Nebula in Lyra. Death of a star comparable in mass to sun. Chemically enriched matter wafted outward, remnant white dwarf core exposed at ring's center. (Palomar Observatory photo)

nuclear power reactor leaves an extremely hot and only slowly cooling core in its wake, so too does the cessation of fusion in a star leave a glowing mass from which heat escapes only gradually. Furthermore, the hottest part of the star is viewed directly, its insulating overlayers having departed as a smoke ring into space. Hence the surface temperatures glimpsed are high enough to merit the descriptive term "white hot." The entire corpse resembles a single gigantic atom, its structure propped up by the same pressures that dictate how an atom's orbital shells will be populated by electrons. Who dares to imagine a more fitting eternity for the celestial object whose steady outpouring of energy permits our existence than as a ten-billion-billion-billion-carat (10^{28}-carat) diamond?

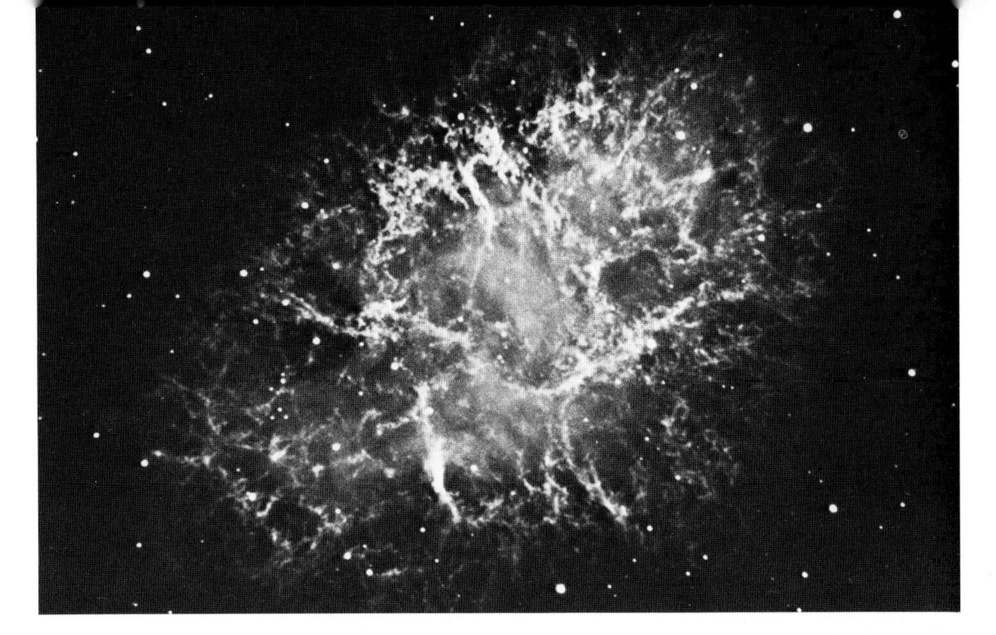

FIGURE 27
Crab Nebula, debris of stellar explosion seen in A.D. 1054. Death of a star several times solar mass. Matter ejected violently. Of two bright stars near nebula's center, one at lower right is core of dead star (see also fig. 28). (Lick Observatory photo)

Many stars, however, enclose several times the mass of the sun. They "live" profligately, burning brighter and faster than their lesser stellar neighbors. Does death extract a heavier price for these more reckless life-styles? Yes in the sense that their final dissolution is more violent; but no in the sense that the remaining corpse is, if anything, more exotic than a diamond. Their greater masses ensure a proportionally greater self-gravitation. Even the internal pressures common to atoms are insufficient to halt collapse. Electrons and protons are squeezed by the vise of gravity into neutrons. For stars a few times solar bulk, a different resisting pressure eventually asserts itself: the assemblage of neutrons resists further compaction, relying upon the same mechanism that dictates its arrangement within atomic nuclei. This last line of defense establishes itself only after the remnant has shrunk to the dimensions of a good-sized city. And within these restricted confines is all the matter of a few suns! Our proverbial teaspoon now contains a hundred million (10^8) tons of matter, perhaps the equivalent of a Golden Gate Bridge. The entire configuration behaves as a single atomic nucleus. The stately rotation it enjoyed in its youth has accelerated to the frenzy of a whirling dervish for the same reason

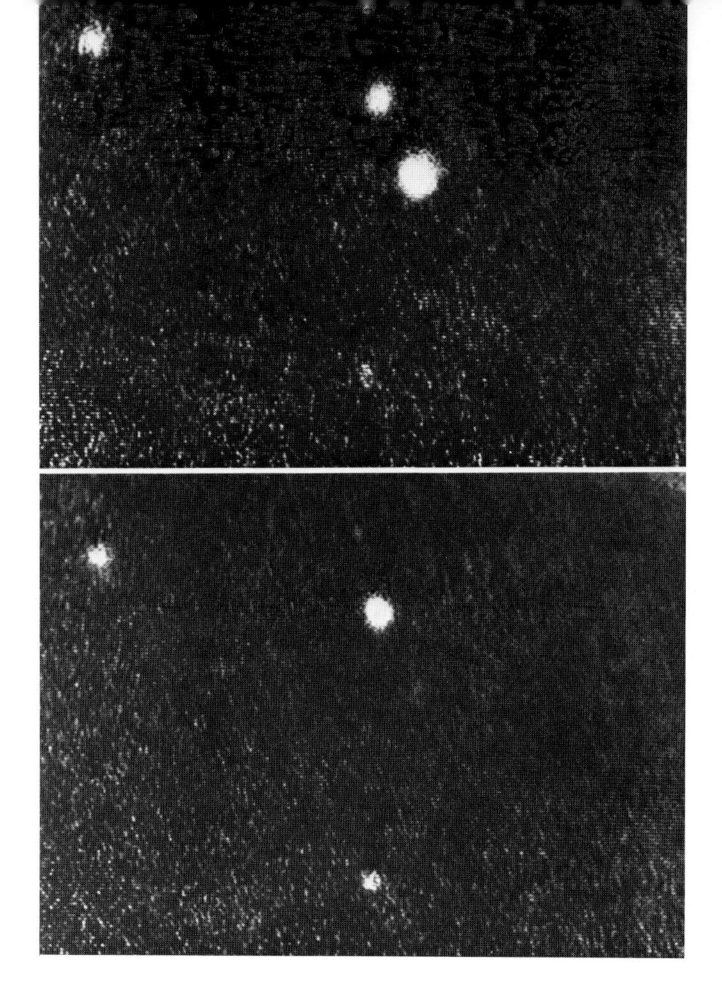

FIGURE 28
Neutron star or pulsar within Crab Nebula (fig. 27). Superdense object winks on (top) and off (bottom) every few thousandths of a second as rotation sweeps bright spot through observer's vision. (Lick Observatory photo)

that a spinning ice skater changes into a blur of motion by drawing in both arms and legs. Likewise, whatever magnetism once existed is intensified many, many times over as the magnet is squeezed ever smaller. The poles of the star become radiating "hotspots," winking at a distant observer each time they sweep across his field of view. These stellar lighthouses are called pulsars for the repetitive character of their signals, or neutron stars for the physical structure they embody. Either way, their reality is stranger than fiction.

FIGURE 29
Relative sizes of sun and stellar corpses.

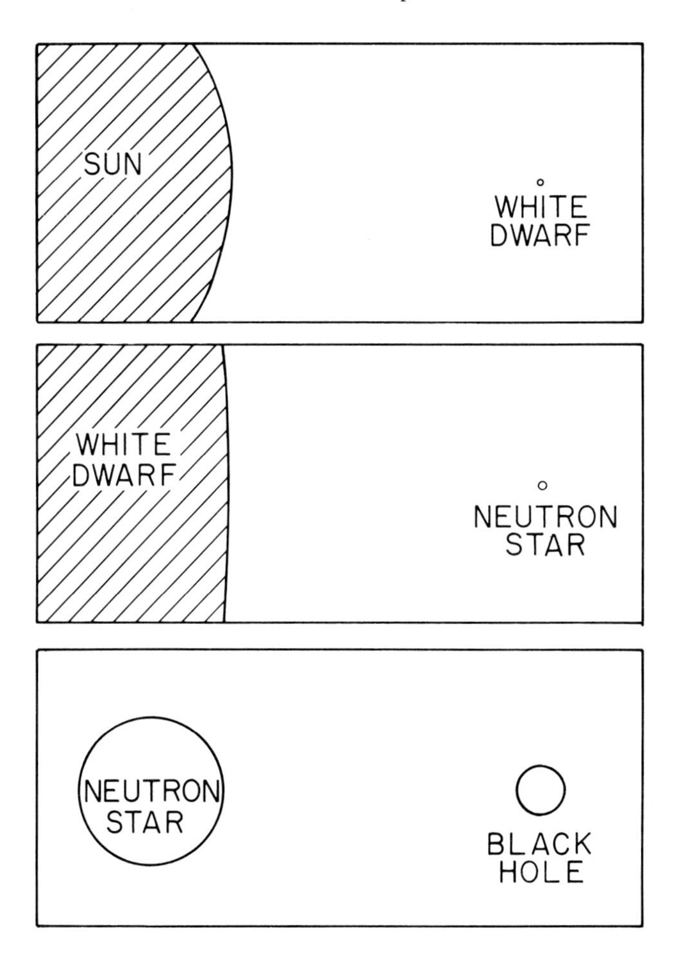

Yet there is more. Those stars an order of magnitude heavier than the sun tolerate no resistance to collapse. Atomic and nuclear pressures are easily surmounted. These leviathans crush themselves out of existence, finally disappearing from sight when the gravitational fields they anchor become so intense that even light is unable to escape the grasp. "Black" accurately describes their resultant countenances, and holes are their contributions to the fabric of space. Their densities are such that a teaspoon of black-hole matter

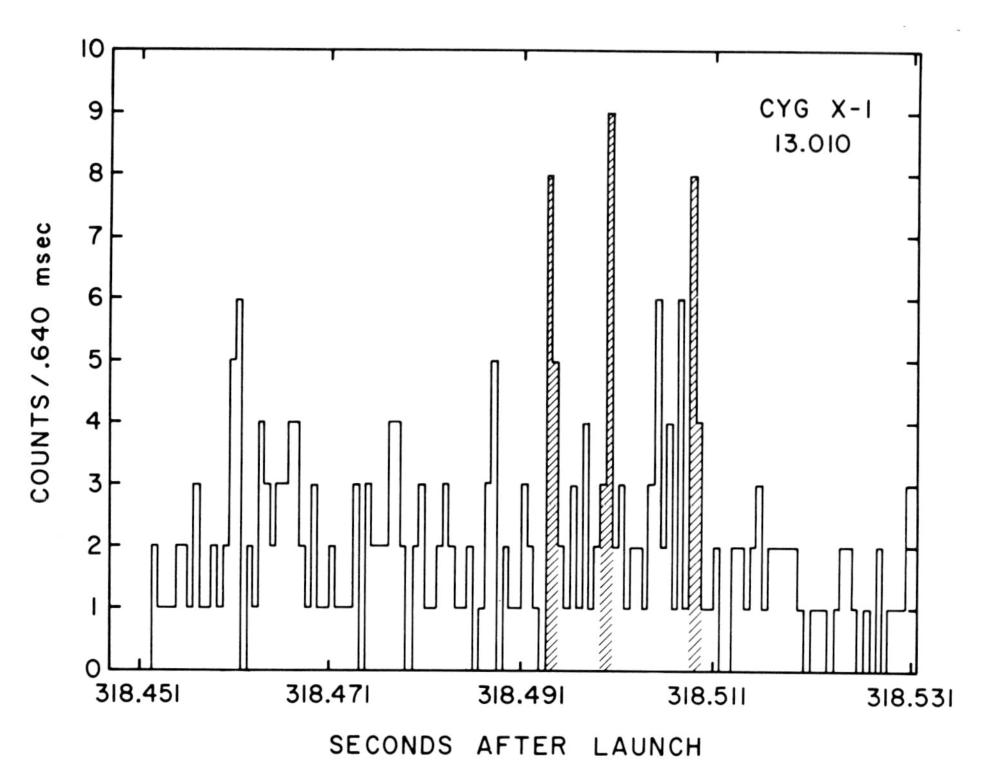

FIGURE 30
X-ray bursts from vicinity of probable black hole called Cygnus X-1. X-ray emission requires high frictional heating, so matter is dense. Hatched spikes endure less than a thousandth of a second, so emitting region is small. (Courtesy R. E. Rothschild, E. A. Boldt, S. S. Holt, P. J. Serlemitsos, and NASA/GSFC)

would weigh as much as Earth itself. Although no signals can return from beyond the brink separating them from the rest of the universe, anything can still fall from this side to that. Since both time and space are warped savagely in the neighborhood of a hole, matter approaching it is swept rather like water into a whirlpool: speeds increase and density rises nearer the vortex. Friction within the matter funneling toward the hole heats it sufficiently that X rays are emitted, characterized by a flickering and fleeting pattern as each successive radiating packet falls over the edge. So, although we can never see black holes directly, we can, and perhaps have, detected their presence through their effects on matter in their neighborhoods.

FIGURE 31
Radio telescopes at the California Institute of Technology's Owens Valley Radio Observatory. Designed for studying molecular concentrations, they also illustrate how much humans' technology has advanced from the stone tools of a few million years ago.

Stars initially condensed within regions of extra material density. It follows that their deaths will likewise occur in crowded surroundings—crowded in an astronomical sense only. The chemically diverse atoms they blow back into space will therefore have a greater-than-average opportunity to meet partners. Atomic alliances constitute molecules, the formation of which is another necessary step in the progression toward living systems. Of course, the likelihood of two atoms meeting in the same place at the same time is exceedingly remote, so sparse are the densities within even the most compact interstellar clouds. But the presence of dust grains within these clouds relaxes the stringency of the simultaneity requirement. One atom can affix itself to a passing grain and await, for millions of years (in a cosmic context, only a fleeting moment), the arrival of another. In addition, the dust shields those molecules that have formed from the most lethal (chemically) rays of starlight.

A close look, this time with radio telescopes because of their penetrating vision, into the dark, dense, dusty regions of interstellar space reveals a varied menu of molecular concoctions (table 1). One detects everything from formaldehyde to smog in the polluted space between the stars. The distribution of molecules is not limited. They broadly populate the Milky Way's many addresses and are found in other galaxies as well. Nor are they all simple; the most complex, cyano-octatetrane (HC_9N) consists of eleven atoms. Most of the detected molecules—those that contain the especially bondable atom carbon—are organic, a name derived from their presence in living organisms. Can it be that the stuff of which we are made, the units from which we are assembled, are scattered hither and yon throughout the universe? That this seems so should, at the least, cause us to consider seriously the possibility that life could arise elsewhere than on the planet with which we are intimately familiar. We—wherein I include salmon, salamanders, and sardines with the *sapiens* among us—can ill exude biological chauvinism.

"From dust to dust" summarizes the life cycles of stars. But the dust dispersed at death has far richer chemical diversity than that consumed at birth. Consequently, these cycles constitute necessary, albeit seemingly unlikely, events in our cosmic genealogies. That we (again, *we* in its broadest sense) are direct descendants from ancestral stellar giants establishes a tie between us and those nightly beacons from which we had heretofore seemed wholly divorced. The iron in our blood, the calcium in our bones, even the gold fillings in our teeth, are sprinklings of stardust manufactured in, and explosively spewed from, physical predecessors that now exist only as galactic memories. If ancestors could in imagination be chosen by their descendants, what flight of fancy would top the reality of being children of the stars?

FOR FURTHER READING

BOK, BART J. 1972 Aug. The birth of stars. *Scientific American* 227(2):48–61.

CALDER, NIGEL. 1972. *Violent universe.* New York: The Viking Press. Pp. 31–80.

JASTROW, ROBERT, and THOMPSON, MALCOLM H. 1974. *Astronomy: fundamentals and frontiers.* 2d ed. New York: John Wiley and Sons, Inc. Pp. 149–195.

PENZIAS, ARNO A. 1980. Nuclear processing and isotopes in the galaxy. *Science* 208:663–669.

SHIPMAN, HARRY L. 1976. *Black holes, quasars, and the universe.* Boston: Houghton Mifflin Co. Pp. 25–117.

TURNER, BARRY E. 1973 March. Interstellar molecules. *Scientific American* 228(3): 51–69.

WALDROP, M. MITCHELL. 1982. The hunter and the starcloud. *Science* 215: 647–650.

6
OTHER WORLDS

Deterministically inclined astronomers are convinced by statistical reasoning that what has happened on the earth must also have happened on planets of stars other than the sun. Biologists, impressed by the inherent improbability of every single step that led to the evolution of man, consider . . . "the prevalence of humanoids" exceedingly improbable.

—ERNST MAYR

To this point the stage has been set for the debut of Planet Earth. Later, all attention will be focused on how and what life emerged thereon. This subsequent intellectual journey could lead to a restrictive parochialism, a solar-system chauvinism, an arrogant anthropocentrism. Better that we first consider whether Earth is the only planet in the universe capable of supporting life.

The answer requires a balance between competing influences. Humans, of course, prefer a status of unchallenged superiority (as, for all we know, do spiders and columbines). But our historical record of accurate self-assessment is dismal. We tend always to overestimate our significance, until the retreat of ignorance before the advance of cumulative knowledge forces us not to. Did not the first civilizations of recorded history call their location the Mediterranean, or middle land? Yet this self-proclaimed centrality vanished when Earth's spherical shape was deduced, a shape whose surface is without center. Further, for how many centuries did humans cling to the Ptolemaic notion that Earth remains the motionless pivot about which all heavenly bodies revolve, before Copernicus dismissed planets to the wings and advanced the sun to center stage? Even the sun proved unworthy of such billing when its location in the outskirts of the Milky Way was revealed. And only in this century did that stellar family come to be perceived as but one among a multitude of equals.

Historically, therefore, preeminence has always been a delusion rooted in ignorance. May it not also be so if we humans presume ourselves the highest form of life in the universe? Or our planet the only one capable of harboring living systems?

But idle speculation knows no limits. Are there not scientific arguments supported by hard data? Are there, for instance, other planetary systems surrounding other stars?* The straightforward answer is that we have no

*The question may be unduly restrictive, for who is to say that life of some type cannot exist in outer atmospheres of stars, or on interstellar grains of dust? In this connection, see especially Feinberg and Shapiro (1980). Nevertheless, for a start we can seek life whose habitat approximates our own.

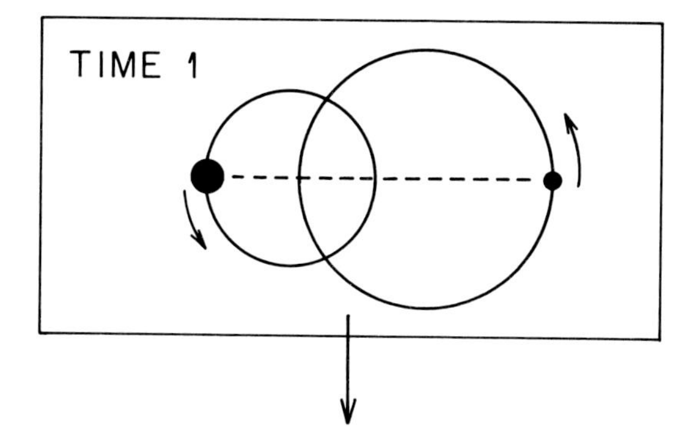

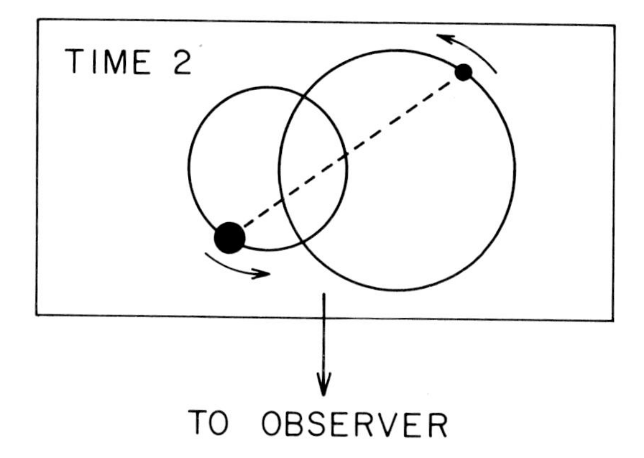

FIGURE 32
Objects orbiting a common center of mass. If either object is invisible, it may still be manifest as wobble in position of its partner.

direct observation of any planet beyond the solar system. Nor have we any technique for finding one directly. The light reflected off planets is simply overwhelmed by the brilliance of their parent stars. In addition, the nearness of parent and offspring means a distant observer sees them as one. The planet Jupiter, for example, when viewed from even the closest star, shines visually with less than a billionth (10^{-9}) the intensity of the sun and is separated from it by only a few seconds of arc (approximately one-thousandth the angular diameter of the sun and moon).

Fortunately, however, there are indirect means for seeking planetary companions. One involves detecting the wobble experienced by every star wedded gravitationally to a companion. Both mutually orbit about their common center of mass. In the case of a star and a planet, as on a seesaw with adult and child, the center of mass lies close to, if not within, the heavier member. Its orbit is consequently small; as the star traverses it only the slightest wobble will be apparent to a distant observer. Yet just such minor perturbations may have been seen in the motions of Barnard's star and a handful of other nearby neighbors. If so, they signal the presence of invisible partners the size of major planets. The measurements demand the highest precision and must be carried out for decades. Unfortunately, different teams of equally careful observers have not agreed, nor is there any prospect of reconciling their measurements short of continuing them for as many years as have been invested already.

Although this first technique offers hope that planets surround stars other than the sun, the controversial evidence cannot be accepted as conclusive. The case for other planetary systems need not rest upon this single shaky reed, however. The universe provides examples of structure on many size scales which point toward the likelihood of objects accompanied by satellites. Recall the tendency for galaxies to cluster, and how within even the most familiar cluster, the Local Group, minor galaxies attend each of the two most dominant members. The very existence of galaxies themselves reveals the gregariousness of stars. Evidently they form only where others have, often in flattened disks rotating about a dense central concentration (fig. 23). But this description of a spiral galaxy applies equally well to the solar system. And that system itself contains subunits repeating the pattern of a relatively massive central body anchoring subsidiaries in flattened surrounding distributions: Jupiter and Saturn, for example, are each accompanied by rings and by more than a dozen moons.

FIGURE 33
The Saturn system, a flattened distribution of rings and satellites surrounding a massive central concentration. (Courtesy NASA/JPL)

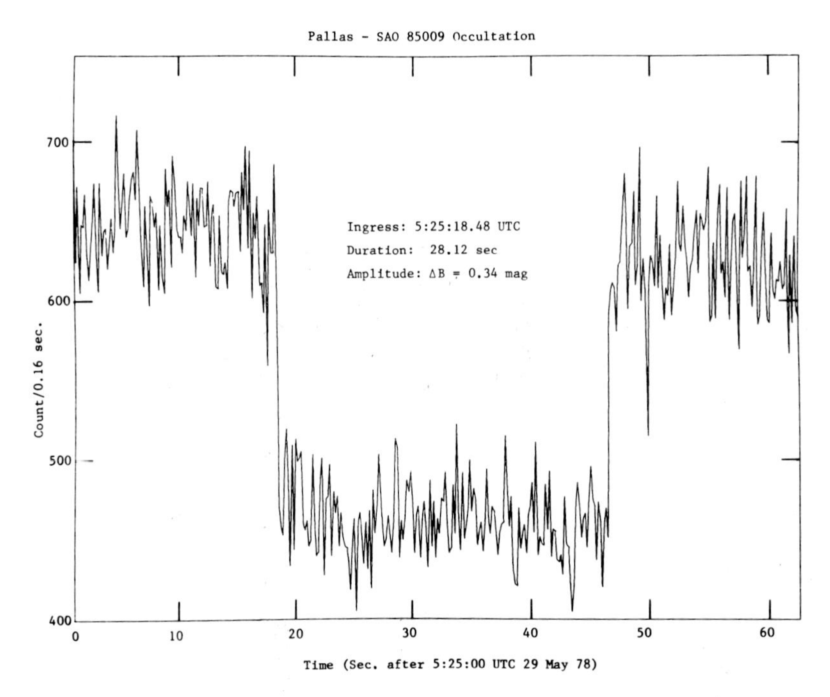

FIGURE 34
Occultation of star by asteroid Pallas and its satellite. Pallas obscures the star from about eighteen to forty-seven seconds. At about fifty seconds, sharp dip indicates passage of satellite in front of star. (Reprinted from Richard A. Kerr, *Science* 211:1333–1336, 1981. Copyright 1981 by the American Association for the Advancement of Science. Original source: Richard Radick, Sacramento Peak Observatory, Association of Universities for Research in Astronomy, Inc.)

The evidence for association can be carried still farther. Tens of thousands of minor planets, called asteroids, orbit the sun in a belt between the paths of Mars and Jupiter. Only a few hundred are larger than sixty or so miles; the summed mass of them all is at best only a fraction of that of the moon. Yet even this insignificant cosmic detritus repeats the pattern established for the very large, some minor planets apparently being escorts for "minor satellites." In fact, the phenomenon may be common. So even though we may never understand precisely how every astronomical system forms, observational evidence indicates that relatively minor objects accompany all major ones. It would seem, therefore, that planets surrounding stars are not unlikely.

Helmut Abt and Saul Levy have carried out a detailed quantitative analysis of stars resembling the sun in color and luminosity, aimed at determining the frequency of companions. Only the 123 nearest were studied because only for them was it certain that present astrophysical techniques could detect stellar companions. In addition to the wobble techniques described previously, three other sensitive tests were applied. These demonstrated that more than half of the stars observed had certain stellar companions, and a logical extrapolation to those remaining implied they, too, were not single. The extrapolation involved comparing the masses of secondary and primary partners. As the ratio of the former to the latter decreased, the number of systems exhibiting that ratio likewise decreased, but always in a steady fashion (see fig. 35). It would be unlikely if at precisely the end of the measured curve—the point where the secondaries became too light compared to the primaries to show up in the various tests conducted—the smooth distribution dropped abruptly to zero. If instead the established relationship continued smoothly into the unexplored realm, then *all* sunlike stars studied have companions. And in 20 percent of the cases the companions are small enough to be planets!* The sun apparently had one chance in five of possessing a supporting entourage. Given all the solarlike stars in the Galaxy, as well as all the galaxies in the universe, winnowing the total to just one in every five still leaves a staggering number of planets where life could abound.

We must temper our optimism somewhat, however, for not all planets will be suitable for supporting life. The evolution of life is decidedly slow, continuing for billions of years in the one sample available for study. Since it is a statistical process, it is unlikely to be significantly faster elsewhere. It follows that the hot, blue supergiant stars, whose lifetimes are measured in mere millions of years, must be excluded as potential parents for thriving planetary biosystems. Furthermore, life exists by virtue of chemical reactions, and

*Abt is more cautious about extrapolating his data to planetary masses. See Billingham (1981).

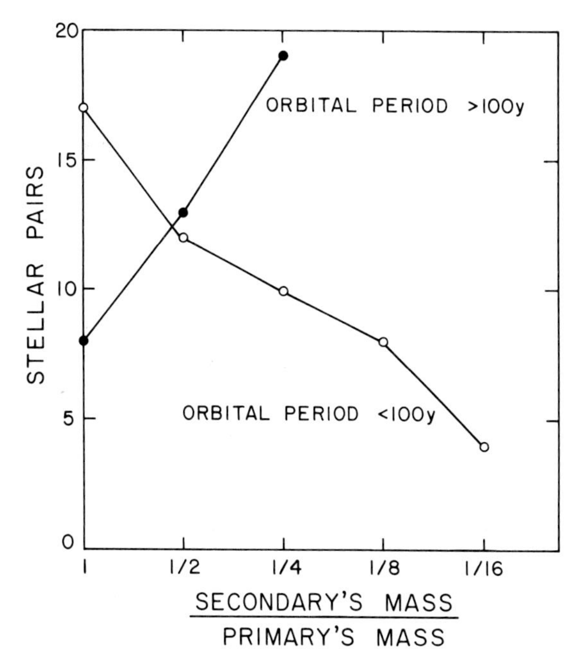

FIGURE 35
Companions of sunlike stars. Stellar pairs sorted according to ratio of masses of two members. Numbers of pairs at each ratio plotted separately for long- and short-period orbiters. If one member of a long-period pair has less than one-fourth the mass of the other, wobble in other member's path is too slow to appear in test conducted. Likewise for short-period orbiters, ratio less than one-sixteenth introduces undetectable wobble. In both cases, however, reasonable extrapolation to smaller ratios predicts non-zero numbers of pairs. All sunlike stars may have companions. (Adapted from Helmut A. Abt, *Scientific American* 236(4):96–104, April 1977)

chemical reactions are notably sensitive to temperature. If too cold, reaction rates are drastically slowed, if not halted altogether; if too hot, the reacting chemicals are themselves destroyed. Life-sustaining planets must therefore be just the proper distance from their parent stars. Of the nine surrounding the sun, for example, only Venus, Earth, and Mars appear to be properly situated. Then, too, life constantly struggles to survive within its environment. Survival is enhanced if environmental extremes are somehow moderated. Since planets in highly elliptical orbits are sometimes far from and sometimes near their

patron energy sources, they endure threatening extremes of hot and cold.

Extreme pessimists can further narrow the available life-supporting options. Perhaps atmospheres provide the seminal chemical setting from which living systems can assemble themselves. Their presence, if Earth is a suitable guide, requires venting of gases from the planet's interior. If so, life-bearing planets must be geologically lively for at least part of their history, possibly exhibiting volcanism and other signs of tectonic activity. Nor can the planet be too light, lest its gravity prove insufficient to retain a gaseous blanket.

A chemically laden atmosphere evenly dispersed about its retaining planet would be a poor setting for embryonic life: the chemicals are simply too widely scattered to interact exuberantly. Far better that they be concentrated, perhaps as on Earth in bodies of water. But whole, vast oceans likewise imply dilute concentrations. Tidepools, on the other hand, offer ideal spawning grounds. Of course, to have tides, a planet must have a satellite, so yet another barrier to the birth of life can be intellectually erected. But what is the point? Our profound ignorance of the full range of conditions under which life can evolve, limited as we are to a partial knowledge of the single sample of which we are a part, warns against imposing too stringent restrictions. The preponderance of evidence suggests there are ample locations in the universe where life could take root.

Of course, places where life *could* survive are necessary, but not in themselves sufficient, to ensure its existence. Potentially fertile soils remain barren unless sprinkled with seeds. But the "molecular seeds" from which entire biosystems can sprout have been spotted in profusion throughout interstellar space. Their formation abiologically must, obviously, be easy. And planets about all but the earliest generation of stars should be endowed with the raw materials necessary for assuring their formation. Moreover, these raw molecules can, in a natural way, be converted into essential building blocks of living systems, as demonstrated repeatedly in laboratory experiments. Whenever mixtures of

glycine alanine valine cystine methionine

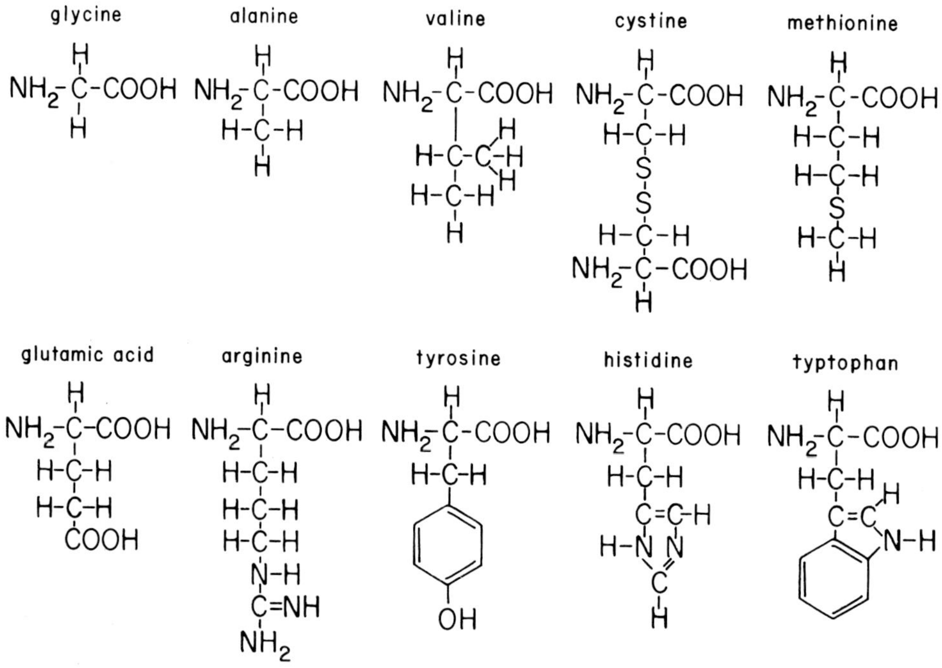

glutamic acid arginine tyrosine histidine typtophan

FIGURE 36
Representative amino acids, fundaments of protein. Note similarity of subunits to interstellar molecules in table 1.

various simple, cosmically popular molecules are infused with energy, the result is always the same: the production of amino acids. And the linkage of amino acids creates proteins, necessary constituents of all living things. The laboratory sources of energy and their natural analogues may be electric sparks, simulating lightning; heat, as from volcanoes; ultraviolet radiation, as would reach the surface of a thinly veiled planet; shock waves, emulating the passage of meteorites; or any number of other natural events. So the actual foundations of life can emerge from chemical reactions among universally present molecules driven by natural energetic events.

The creation of amino acids on Earth could not have been so extraordinarily difficult as to have occurred only in this one location. Of that we can be certain because of messages brought to us by frequent extraterrestrial visitors, meteorites. These fragmentary leftovers of the primordial solar nebula have been painstakingly analyzed chemically. Among their ingredients are amino acids, whose extraterrestrial origin cannot be doubted because the structures of half of them are mirror images (right-handed instead of left-handed) of the only kind present in the plants and animals of Earth. Although life has yet to be detected beyond our terrestrial environs, its immediate chemical progenitors are almost certainly widespread and abundant.

In any discussion concerning life in the universe, we should examine closely the one sample of whose existence we are certain. What characterizes life on this planet? In particular, are there any properties that enhance the prospects of its existence elsewhere?

Terrestrial life's most paradoxical property is the commonality of its biochemical basis contrasted with the amazing diversity of individual forms. What could superficially seem less similar than a mole and a rose, or an alga and a hippopotamus? Yet each is linked genetically with every other, as indeed with every individual that ever has lived or ever will live. Their chemical structures are stunningly alike, all built from the same pool of twenty amino acids ever so slightly rearranged. The number of amino-acid rearrangements indicates the evolutionary separations. But it is in the prodigal diversity of living forms, not in their biochemical sameness, that the strength of Earth's biosphere resides. The planet's environment has been frequently altered—by earthquakes, ice ages, sliding continents, supernova explosions, volcanoes, asteroid impacts, and even by emerging life forms themselves (especially those who read and write these words). Each alteration tests the system's resiliency. Most species have failed their tests: that is, for every one in existence today, thousands more have long been extinct. But at each geologic or biological crisis, because of the wide range of types and forms, some have always survived to continue the progression.

Of course, diversity is not a mere accident. The slim membrane of gas and water enveloping Earth, within which all living species reside, is neither

FIGURE 37
Antennariid anglerfish. Lurelike appendage represents ingenious adaptation for capture of energy. (From T. W. Pietsch and D. B. Grobecker, *Science* 201:369–370, 1978. Copyright 1978 by the American Association for the Advancement of Science.)

uniform nor homogeneous. Some regions are hot, others cold; some wet, others dry; some acidic, others alkaline; some fertile, others sterile. Each niche imposes its own requirements upon what can live within. No one expects to see a jellyfish on a mountain summit or a bighorn sheep on an ocean beach. Yet one does expect to see some form of life in each location. Obviously there are many opportunities for varying life-styles. And because terrestrial life has proved adaptable and flexible, equally many opportunists have evolved.

In short, there are several ways to make a living; conversely there are few options that have not been naturally selected. Some of the adaptive options can only be described as ingenious. An example is the antennariid anglerfish. This miniature Izaak Walton provocatively dangles an appendage resembling nothing so much as a small fish. The unsuspecting predator who succumbs to the allure of this bait will not, however, find a meal, but become one. By dint of its innovative deception, the small antennariid has developed a brilliant strategy for survival.

Many organisms have adapted to environmental extremes. We know life exists on the highest mountaintops and in the deepest ocean trenches, in the hottest deserts and on the polar ice caps. Some life-bearing environments can only be described as hostile, perhaps none more so than the boiling, acidic springs of Yellowstone National Park. But the tenacity and determination of life are such that even the dual barriers of these springs are not insuperable: within thrive bacteria named, horribly, *Sulfolobus acidcaldarius*, whose lifestyles, though seeming alien in today's world, may have been mainstream on primitive Earth. In fact, these bacteria and a few close relatives have probably been present since the genesis of life on this planet, a record of endurance which should give pause to the self-heralded superiority of humans.

The diversity, adaptability, and tenacity of earthly life have favorable implications for its existence elsewhere. Clearly, we do not know what boundaries specify the limits of life even on this planet. We should therefore never be too quick to dismiss any cosmic locale as inhospitable—worse yet, hostile—to the presence of life. In our search for its existence, we must never underestimate the determination, the dogged persistence, of living systems. Of course, we shall find nothing resembling life as we know it. Our sample has been finely honed to fit the specific requirements of the third planet from the sun. Every other potential environment will be equally unique. The succession of biological experiments posed by genetic mutations will have different outcomes in different settings. Since their occurrence is pure happenstance anyway, their outcome here on Earth could never be precisely repeated. Hence the diversity of earthbound life about which we marvel represents but a narrow slice of the total possible spectrum. Perhaps a near-infinity of living creatures, some wilder than we are capable of imagining, await our discovery—or we theirs.

FIGURE 38
Bristlecone pine tree, example of life's tenacity. Resident at 10,000-foot elevations in dry climates swept by high winds and numbing cold, sprouting from soil too poor to host competing plant life, these specimens persist longer than any other living systems.

To appreciate another property of earthly life we should escape the confinements of membership within and retreat once more to a cosmic perspective. From the depths of space we can see that the biosphere pulses and throbs as a single unit. No species can exist without many others; all are dependent upon the integrity of the whole. But mutually interdependent and interacting systems creating a whole greater than the sum of its parts are not without precedent. Consider, for example, the complex that is each one of us. Although single organisms only, we operate by virtue of the cooperative and coordinated functioning of several interrelated subsystems—skeletal, muscular, circulatory, nervous, digestive, reproductive, respiratory, and so on. In turn, each subsystem is an immense collection of specialized cells molded into various internal organs—a heart, lung, pancreas, biceps muscle, whatever. And within every cell are structures resembling the most primitive protozoans. Systems within systems within systems, none self-sufficient, all essential to the operation of the whole: does this not describe the biosphere as well? If so, it is alive, an organism within which we are mere cogs, neither superior nor inferior biologically to our coplaneteers.

All species (or cogs) have a characteristic without which none could be separably identifiable, namely, a continuity of form. Although individuals come and go, a pattern and structure persists in their descendants, in their descendants' descendants, and in all who follow thereafter. No individual is indispensable, but every species is a branch of a "family tree" whose roots extend beyond the genesis of life on this planet and whose subsequent development demands the greatest care. The tree of life need never cease growing, but its future health depends upon its present branches retaining their vitality. The human branch can be particularly shortsighted. Although we alone can deduce the structure of the integral whole, we often fail to act as mere parts of it. This failure could be lethal. Can we deny to posterity the same munificent conditions that attended our own introduction as a species?

Extrapolations from statistics of a single sample have infinite errors. The prospect of adding to the known sample of biohabitats, and thereby reducing our uncertainties about life's boundaries, has lured man to neighboring sites. The results have been either disappointing or ambiguous. Not unexpectedly the moon—with its dry, fiercely pounded surface nakedly exposed to the sun's radiation—was sterile. The stark contrast between vibrant Earth and inanimate moon, so evident when both are seen at once from afar (fig. 5), speaks powerfully to our consciences: the life that so distinguishes Earth occupies only a thin veneer over a base as desolate as the lunar surface. Those who would harm or destroy the preciousness of our setting should seriously consider the alternative.

Mars, in contrast with the moon, quickened our expectations. Although the romanticism of a civilization of canal builders had vanished upon close examination, it was replaced by the reality of spectacular canyons sculptured by once freely flowing water (fig. 6). Martian seasons complete with the advance and recession of polar ice caps, as well as a changing planetary countenance hinting at variations in vegetation, suggested the existence of life. The central mission of the Viking spacecraft was to confirm or deny our hopeful expectations. The outcome, however, was to leave us as uncertain as before the quest.

The Viking landers carried apparatus to analyze the Martian soil in four ways. All four tested for telltale by-products of the metabolic processes of Earthlike life.* For example, animals, ourselves included, expire carbon dioxide, and plants, oxygen. A natural question, then, asks whether the Martian air is altered by organisms buried in that planet's soil. A small sample was warmed and humidified, giving any dormant seeds or spores a chance to spring to life. Initial results indicated a burst of oxygen was indeed released,

*Critics contend that the experiments were poorly designed for this very reason: Mars, after all, presents niches differing greatly from Earth's.

exactly as anticipated if plants resided within. The same result, however, could be expected if the soil contained some chemically unstable compounds, quick to free the oxygen which they only loosely incorporated. As a result, this first experiment, however tantalizing to exobiologists, cannot be used as unequivocal evidence for the presence of life.

Are there microbes, food consumers, whose appetites might reveal their existence? To find out, another soil sample was drenched with a nutrient broth, almost the proverbial "chicken soup" that cures all ills. If organisms converted this food into energy and tissue, they would release waste gases, most prominent among them being carbon dioxide. To ensure that the gas was released as a result of animal digestion, the food was "tagged" by building into it a radioactive form of carbon. Sure enough, considerable radioactive carbon dioxide was given off by the treated soil. Animals? Possibly, but again chemists were able to concoct an alternate inorganic explanation.

Well, then, is food manufactured by microorganisms in Martian soil? We know, for instance, that plants on Earth photosynthesize—that is, meld carbon dioxide, water, and energy drawn from their surroundings into carbohydrates. Why not illuminate a Martian soil sample exposed to an appropriate (for Earth) atmosphere? After several days' incubation, one might expect any plants to have incorporated the atmospheric carbon, again a radioactive variety for traceability, into their structures. They may, in fact, have done so, for the soil was found to be highly radioactive. As a further check, the artificial sunlight was left off, and the radioactivity in the soil dropped correspondingly. Chemical explanations, although forthcoming, all appeared rather contrived. The manufacture of food by plants seemed indicated.

But a fourth test contradicted the optimistic findings of the third. Every organism on Earth consists of organic molecules. Their presence is found in every locale searched: Antarctica, downtown Manhattan, Mount St. Helens, and the Gobi Desert. Yet a detailed chemical analysis of Martian soil yielded no organic molecules. Their absence seemed to preclude the existence of living organisms.

FIGURE 39
Ghost town of Bodie, California. Humans today can extinguish life on Earth, leaving a ghost planet for all eternity. The prospect is especially tragic if Earth is the only oasis in the universe.

The last two tests cancel each other, while the first two are ambiguous. We are left exactly where we were before the Viking missions. We have no conclusive evidence for the existence of life on Mars but no absolute proof of its absence.

Exobiology is still a scientific discipline with no data. But its motto remains "absence of evidence is not evidence of absence." Since humans are the first on this planet to acquire an awareness of its setting in time and space, we bear the burden of searching for cosmic kinships. If we are alone, the burden is especially onerous; for upon us falls the responsibility of sustaining and expanding the only oasis—the extinction of which would make the solar system a ghost town whose funereal presence would haunt eternity with a reminder of what might have been. If, instead, we are a part of a much larger cosmic community, our participation in its furtherance cannot be postponed indefinitely.

FOR FURTHER READING

ABT, HELMUT A. 1977 April. The companions of sunlike stars. *Scientific American* 236(4):96–104.

BILLINGHAM, JOHN, ed. 1981. *Life in the universe*. Cambridge, Mass.: The MIT Press. Pp. 203–205.

BINZEL, R. P., and VAN FLANDERN, T. C. 1979. Minor planets: the discovery of minor satellites. *Science* 203:903–905.

BRACEWELL, RONALD W. 1975. *The galactic club*. San Francisco: W. H.Freeman and Company. 141 pp.

FEINBERG, GERALD, and SHAPIRO, ROBERT. 1980. *Life beyond Earth*. New York: William Morrow and Company, Inc. 464 pp.

GORE, RICK. 1977. Sifting for life in the sands of Mars. *National Geographic* 151(1): 9–31.

KERR, RICHARD A. 1981. Satellites of asteroids coming into vogue. *Science* 211: 1333–1336.

JASTROW, ROBERT. 1977 March. Report from Mars. *Natural History* 86(3):48–53.

MILLER, STANLEY L., and UREY, HAROLD C. 1959. Organic compound synthesis on the primitive earth. *Science* 130:245–251.

MOSSER, JERRY L., and BROCK, THOMAS D. 1979. Taking the heat. *Natural History* 88(2):84–91.

O'LEARY, BRIAN. 1980. Searching for other planetary systems. *Sky and Telescope* 60:111–113.

OVENDEN, MICHAEL W. 1962. *Life in the universe*. New York: Doubleday & Company, Inc. (Anchor Books). 160 pp.

PIETSCH, THEODORE W., and GROBECKER, DAVID B. 1978. The compleat angler: aggressive mimickry in an Antennariid anglerfish. *Science* 201:369–370.

THOMAS, LEWIS. 1975. *The lives of a cell*. New York: Bantam Books, Inc. 180 pp.

7
BIOLOGICAL EVOLUTION

So, thought Herzog, acknowledging that his imagination of the universe was elementary, the novae bursting and the worlds coming into being, the invisible magnetic spokes by means of which bodies kept one another in orbit. Astronomers made it all sound as though the gases were shaken up inside a flask. Then after many billions of years, light-years, this childlike but far from innocent creature, a straw hat on his head, and a heart in his breast, part pure, part wicked, who would try to form his own shaky picture of this magnificent web.

—SAUL BELLOW

Throughout the whole of cosmic history one message rings clearly: the universe is dynamic, manifesting relentless change. From the moment of instantaneous creation of matter and energy, there have followed aeons of its rearrangement and transformation: first into galaxies, then into stars within, the latter synthesizing atomic raw materials into ever richer chemical combinations. Nothing in this history suggests that this late-blooming planet will be static either, and indeed it is not. The changes that are wrought in the natural endowment of Planet Earth are, in fact, particularly wondrous, for through them matter assembles itself into forms that resist the natural tendency toward dispersal into chaotic randomness. Instead, the tendency is toward order, complexity, diversity, uniqueness. In short, the matter is alive.

The pace at which living matter has blossomed has been sporadic. Relatively long intervals of apparent quiescence and stability have been frequently interrupted by short epochs of tumultuous change. The whole story spans 4-1/2 billion years, an interval of such duration as to defy comprehension. In order to display the pace in an intelligible fashion, imagine the totality of Earth's history compressed into a single year beginning with New Year's Day and ending now, midnight on December 31. Each week of this imaginary year spans 87 million (87×10^6) actual years; each day, 12 million (12×10^6). Every hour represents the passage of half a million (500,000) years; every minute, 8,500 years; and every second, 140 years. Initially we can leap months and weeks at a time, but by year's end, significant changes occur in fractions of a second.

The driving force behind the process of continuous change is reckless opportunism. Nature seeks stability and permanence, but it approaches these conditions by generating prodigal variety. Witness, for example, the uniqueness of each individual fellow human. With the possible exception of identical twins, there is some characteristic or combination of characteristics that renders each of us identifiable; so, too, with spiders, coyotes, and Joshua trees.

Some of these distinguishing characteristics will be more favorable than others in coping with the circumstances of that individual's existence. Those who have these characteristics will thrive and produce more like offspring than will less suited individuals. The set of favored characteristics will therefore preferentially survive the filtering or screening established by the environment. In this way, the fit are winnowed out from the marginal or the unfit, defining a stable configuration for as long as the environmental dictates remain unchanged. We shall witness this process played over and over during our passage through the imaginary year, concentrating on those adaptations that proved so advantageous to their possessors that they originated new life forms.

Of course, each new organism becomes part of the environment of every other organism. All interact vigorously in many ways—as predators, competitors, hosts, and habitats, for example. Life's effect on the environment is therefore as profound as the environment's effect on life. The two, inhabitant and habitat, cannot be separated. Traditionally biologists have interpreted evolution in terms either of adaptations to changes in the natural setting or of novel strategies for exploiting existing opportunities. There is, however, another viewpoint, one more attuned to a cosmic perspective, which reverses this reasoning. Life is no longer the passive reactor to external events. In fact, life is so intimate a part of this planet that divisions into animate and inanimate are artificial. Instead, life *in toto* actively controls and regulates the operation of the overall system in which it is immersed so as to enhance its own survival prospects. The compressed history that follows will adopt the traditional viewpoint, but we shall return to the so-called Gaia hypothesis when assessing the human condition in the next chapter.

January and February. The recently formed planet Earth is initially molten. Its spherical gravity pulls the heavier matter, rich in iron and nickel, to its core, while the lighter material, soon to be basalts and granites, floats to the surface. Throughout these first two months, the planet is cooling and shrinking. A thin crustal covering begins to jell.

Volcanoes belch gases, including water vapor, from the interior, creating a primitive atmosphere. By today's standards, this atmosphere is not benign: it contains no free oxygen but instead is laden with noxious molecules forged mainly from hydrogen, carbon, oxygen, and nitrogen. Unlike at present, the original atmosphere cannot screen the sun's ultraviolet radiation. Therefore, ample energy penetrates to drive chemical reactions among the accumulating molecular inventory. Atmospheric lightning also stimulates the chemical pot. Among the agents produced are amino acids, building blocks of protein molecules. Since these—the conditions that fostered life originally—no longer exist, we could not again set in motion the blossoming of the planet.

The steam vented skyward cools until water droplets condense. These rain to the surface that, by early February, has cooled sufficiently so that the falling water is no longer boiled off. Instead, it collects in vast oceans. The many molecules randomly combining turn these seas into an organic broth of amino acids, nucleic acids, and carbohydrates. The very ultraviolet energy that sparked the reactions creating these products could also tear them asunder; fortunately, the seas shield the newly formed protobionts from this danger.

Within the seas, especially where concentrated into tidal pools, molecules haphazardly encounter one another. Some have natural electrical affinities for others and hence combine into more complex chemical species. Many long molecules show natural tendencies to curl upon themselves. These segregate from their aqueous solution into microspheres. Such protocells solve a problem that plagues any organism sharing an environment with other organisms: namely, how to avoid being diluted out of existence. Nearly eleven months later, during the last second of this compressed year, fossil bubbles of these ancestors will be found in rock samples.

Among the various products of the ongoing transformation and combination of chemical units will be some that acquire the ability to assemble their likenesses. Their structures may be just such as to serve as templates upon which fragments can attach until complete second units are manufactured. The Siamese twins may then split, each seeking the substance to double again its numbers. The process need be no more mysterious than that by which a salt crystal in saline solution grows, each atom of sodium attracting from its surroundings one of chlorine, and vice versa. Obviously, these replicating molecules will soon outnumber all others.

At some point, late in February, the multiplying and cellular entities will be said to be alive. In fact, the breath of life may stir inanimate matter of different structures in different places at different times. The acquisition of life guarantees a form of permanence, of stability, for living cells have the two central talents of metabolism and reproduction. Metabolism assures short-term survival: the cells can rearrange the atoms and molecules they ingest into those needed for their own maintenance. Reproduction assures long-term survival by continually populating the environment with likenesses.

The most striking aspect of the origin of life is that it occurred so rapidly—in relative terms, in less than two months out of a year-long history. Because life evidently arose as quickly as possible, it is unlikely to be a complex accident requiring aeons of time to convert the improbable to the certain. Its birth may instead have been as natural and inevitable as the production of quartz or feldspar. Life may therefore be intrinsic to more planets of our size, composition, and position than we have hitherto imagined.

March. By the first of March, some continents have formed. The real interest is not geologic, however, but biological. All the primitive unicells are consumers of organic materials, not producers. But the conditions that provided for abiotic production of organic matter have ceased to exist as Earth has aged. There is therefore an intense competition for food; the first Malthusian shadow casts its pall. Nature finds a timely solution, however: photosynthesis, the inverse of respiration. Blue-green algae and primitive bacteria emerge with the ability to lock captured solar energy into synthesized organic molecules. The first biological production of food has occurred, an event to which all subsequent fauna owe their existence. The original photosynthetic life splits the carbon from atmospheric carbon dioxide and the hydrogen from water, then fuses the two into carbohydrates. The waste product, oxygen, escapes to the atmosphere, initiating a transformation there of profound significance. Life has begun to change the global environment in such a way as to influence all subsequent evolution.

April, May, June. The planet acquires a greenish tint as algae and bacteria diversify, multiply, and disperse. There are no natural enemies or predators to halt their advance across the seas.

July and August. The addition of oxygen to the atmosphere continues. The adaptive range of simple, living cells extends from those intolerant to oxygen, through those that are tolerant but can do without, to those that must have it.

September. The oxygen atoms accumulating in the atmosphere combine into molecules called ozone. The ozone layer shields the planet below from the sun's destructive ultraviolet radiation. Heretofore, life has necessarily sought refuge in the sea. Now, the land itself becomes a safe haven. Life has so modified its environment as to make the latter more hospitable.

Nearly three-fourths of Earth's history has transpired. Yet the only living presence consists of single cells of the simplest kind. The step from inanimate to animate matter was easier than expected; the one from single-celled to multicellular organisms is evidently more difficult.

The biosphere, however, is poised for an eruption. The event that triggers it is the invention of sexual reproduction. For the entire history of the planet to this moment, cells have reproduced asexually through binary fission. A living unit, whose interior was undifferentiated, ingested enough from its surroundings to outgrow its containment mechanism. The result of such overindulgence was a split—a good way to make additional identical units, but a poor way to make different ones. Some small, primitive, nonnucleated cells, however, had probably taken up residence inside larger ones—to the mutual advantage of both. A kind of cell was evolving with specialized internal structures. Each such cell contained only half the genetic material needed to determine its successors. Reproduction consequently required two parents, each contributing half the makeup of its offspring, who would therefore be an exact copy of neither. This is an effective way not only to make more cells but to make different ones as well. The differences between generations can become great if the transmission of genetic information is flawed. The resulting mutations sprout new branches of life's evolutionary development.

October. Multicelled invertebrates flourish. Various worms, jellyfish, and soft corals populate the seas. At last, animals outgrow their previously microscopic dimensions. Multicellular structure confers several advantages, all attributable ultimately to the repetition of cell machinery. For example, longer lives are likely, since backup cells can be substituted for malfunctioning ones. Further, reproduction should be more prolific, because many cells can be devoted exclusively to that simple task. In general, cells dedicated to specialized functions should assure greater efficiency. Finally, the fact of larger size is itself advantageous in that it guarantees a greater stability of internal physiology.

November 1–15. The opportunities incumbent in multicellular construction are seized. Life explodes, but still only within the seas. The land remains barren except for a few rock-clinging lichens. The seas are shallow, the continents low-lying and small. Earth is tranquil, its mood one of agelessness.

Sponges, worms, and related pioneer mollusks prosper during these gentle times. Meanwhile, some animals don suits of armor, hard parts beginning to protect soft bodies. The most successful of these is the trilobite, a cross between today's lobster and sea slug. The world has never seen such an advanced and efficient life form. The dual advantages of locomotion—its spindly legs propelling it across sea-bottom ooze—and protection so overwhelm the competition that nearly three-fifths of all multicelled creatures are trilobites.

November 15–30. The geologic calmness is shattered. Shifting continents pinch off a proto-Atlantic Ocean. Their collision thrusts island arcs and undersea continental shelves skyward. The Appalachian Mountains, as well as the highlands of Scotland and Scandinavia, are created. Many volcanoes are spawned, scattering ash profusely.

Ancestral squid and octupuses emerge, as do clams, starfish, and corals. At the very end of this imaginary year, a creature will reconstruct these happenings. His success in doing so will delude him into assertions of superiority over all other life forms. But what are nature's real criteria for success? The most important is certainly the ability to adapt and survive. Many of the creatures who appear now in late November will be with us as the year closes. They are therefore demonstrable successes at playing the game of life. Only after our own longevity matches theirs can our self-ascendancy legitimately be proclaimed.

The pace quickens again immediately after Thanksgiving. The first vertebrates appear in the form of jawless fishes, ancestors of today's lampreys and hagfish. That great unexploited niche, the land area, is finally invaded: some scorpions and millipedes climb aboard, there to be joined by primitive plants, devoid of roots and leaves, which conduct nutrients and water through what is scarcely more than a bundle of tubes. Has anyone imagined that our place of residence came to be occupied so recently? The appearance of the continents is very rapidly transformed as lush, green, giant tree ferns and early evergreens exploit the virgin territory. In the meantime, diversification continues in the seas. Jaws develop among fishes, greatly expanding their food opportunities. And fins appear in pairs, the resulting improved balance making for better swimmers. Sharks crown this, the Age of Fishes.

December 1–7. The shallow seas advance and recede in response to movements of Earth's crust. Sea-bound life is threatened by the cycles of boom and bust. An adaptation is called for—and appears. Crossopterygii, a group of bony fish, develop both internal air bladders and stout, muscular fins. Given the abilities to gulp air and to walk on stumps in tidal flats, they become bridges between fish and amphibians. Insects, including spiders, debut. No class of life will find the world so much to its liking; by the end of the year, their varieties will constitute three-fourths of the animal kingdom. The tremendous development of plant life continues on land. Rich, lush forests spread abundantly. At the continents' edges these forests are repeatedly submerged by the changes in sea level. Sediments slowly bury them, and the process of organic decay sets in. In this way, fossil fuels are created for the inhabitants who will appear during the year's waning moments. Months in production, these fuels will be totally consumed in mere seconds.

Amphibians replace fishes as the dominant fauna. So much potential food burdens the land that these amphibians are only the first to reap its bounty. But they exploit the land niche only imperfectly. They must return to water to mate and to lay their fragile eggs. The newborn must then pass their adolescence in an aquatic stage. On December 6, final liberation from the bondage of the sea is accomplished by reptiles. The tough-skinned reptilian eggs encase an aqueous environment providing an initial food supply. They can be deposited anywhere, not just at water's edge. Soon, this advantage will be so great that other faunal types will be outnumbered.

December 8–15. The week, the fiftieth of Earth's condensed historical year, begins with a violence to which the planet has become unaccustomed. Mountain building increases in intensity. The climate changes drastically, with glaciers covering much of the land mass and locking up sufficient water to create vast deserts elsewhere. These dramatic changes herald one of the most inhospitable periods for life ever known.

Denizens of fetid swamps are poorly adapted for dryness and cold. As a result, they suffer major extinctions. Fully half the families of marine invertebrates vanish in a fraction of a compressed day. Trilobites disappear forever, ending a remarkably successful reign. Amphibians decline in numbers, although new life-styles—for example, frogs—continue to be introduced. Reptiles prove most adaptive to the transition and radiate prolifically. Turtles, for instance, doggedly continue as in Aesop's fable. They will patiently witness the rise and fall of dinosaurs and the triumphant spread of mammals, and they will still persist until the end of the year, leaving behind the many metaphorical "hares" who run more rapidly but less persistently in the race for survival. Floral life, too, must adapt or perish. Its strategy for dealing with the harshness of the surroundings is ingenious: lock up the makings of a new plant in a tough encasement, a seed. That way, the plants can remain dormant during times of stress, but they can grow rapidly on stored food when conditions improve.

Toward the end of the week, the continents start merging together. By the fifteenth, they are united into a single land mass called Pangaea, while all the water fills a universal ocean, Panthalassa, a remnant of which is our present-day Pacific. Also on the fifteenth, the Age of Reptiles introduces its crowning achievements, the dinosaurs, the largest land animals the world will ever know.

December 16–24. As the year winds down, the landscape remains unfamiliar to those of us who will be present at its end, a testimony to the stunning rapidity with which new opportunity, followed by natural selection, can drive the pace of change. In North America, the upthrust of a range of mountains extending from southern California to Alaska introduces several new niches.

FIGURE 40
Sierra Nevada. Seemingly the result of ancient and slow geologic processes, these mountains date back only two weeks if Earth's history is condensed to one year.

Dinosaurs invade the land (diplodocus, stegosaurus, and others), the sea (ichthyosaurus, plesiosaurus), and the air (pterosaurus). Their immense size may contribute to their success. As they grow ever larger, their volumes increase faster than their surfaces, through which all energy transfer occurs. An internal thermal stability, hence constant metabolism, may result. These giants require an enormous food supply, one that fortunately is available in lush, widespread tropical jungles. Such forests are dominated by floral giants the equal of their faunal counterparts: ancestors of today's sequoias tower over tall, delicate tree ferns and palmlike cycads. Their size and broad geographic distribution cannot compensate for a wearisome monotony, however. Everywhere, Earth is green.

At last, on December 20, flowering plants appear, thereby shattering the monopoly of color. Flowers and insects, whose development has been explosive, immediately enter a marriage—more properly, a symbiosis—which continually adjusts itself to the advantage of both partners. Their refined interaction is well illustrated by the presence of carotenoids in the petals and pollens of many flowers. These vitamins sharpen the vision of their insect consumers in exactly those colors which the flower is offering as an attractant. In addition, flowers send forth molecules, pheromones, whose scent mimics the sexual advertisement of several insects. In return for nourishing insects, flowers benefit from a less haphazard system of fertilization. Soon, many modern trees develop, among them birch, elm, oak, and maple.

Meanwhile, animal life discovers new avenues of opportunity. Because all animals have been cold-blooded, certain times of day are prohibitively hot for them, the animals suffering heat prostration; or excessively cold, causing total inertness. On the seventeenth, mammals inaugurate a new era. By gasping or sweating, they can cope with heat; with hair and fat for insulation and high rates of metabolism for continued production of energy, they can cope with cold. The noontime and midnight niches avail themselves to warm-blooded, mammalian exploitation. Only two days later, another oppor-

FIGURE 41
Flowers shatter green monotony. Symbiosis with insects illustrates how thoroughly life relies upon and provides for other life.

tunity is seized when birds soar skyward. Their three-dimensional locomotion enables them to exploit food sources unavailable to most amphibians, reptiles, and mammals; to escape landlocked predators; and to flee from harsh winters.

Even the land continues its restlessness. Pangaea splits into northern (Laurasia) and southern (Gondwanaland) halves, separated by an embryonic Mediterranean Sea. India and Australia-Antarctica subsequently fragment from Gondwanaland. In what is now North America the crust is bent downward into a trough running from the present Arctic Ocean to the Gulf of Mexico, permitting the sea to inundate that continent for the last time.

The Last Week of the Year. By Christmas Day the world is blessed with an immense and growing diversity of small, primitive mammals and water-shore birds. Reptiles are poised for a fall from dominance. Forests wane before the advancing grasslands. Yet nothing portends the calamity of the following day.

On December 26 the great, thundering dinosaurs vanish forever. When their fossils are unearthed a few days hence, their finders will be electrified by the grandness of dimension, saddened by its disappearance, and puzzled about its cause. Perhaps the entire planet suffered an infrequent, but devastating, natural cosmic catastrophe. An asteroid may have struck it, pounding a sizable crater into Earth's surface and throwing its own and the crater's pulverized contents into the atmosphere. This huge showering of dust would quickly spread worldwide, preventing sunlight from penetrating to the surface below. As a result, photosynthesis would cease, depriving animals of their ready supply of stored energy. If this hypothesis is correct, one expects the clays laid down at this epoch to be rich in those ingredients—specifically iridium and osmium—which are cosmically abundant but terrestrially rare; and so they are found to be. Life on Island Earth evidently cannot escape the harsh realities lurking in its surrounding oceans of space.

Other changes remold the planet's countenance. Europe splits from North America, Africa from South America. The Atlantic Ocean spills into the void. India continues its rapid race toward Asia, foreboding a collision so violent that the world's highest mountains, the Himalayas, will result. Elsewhere, the Rocky Mountains begin to fold noticeably upward on December 27. This large-scale upswelling looses the might of running water seeking the level of the sea. From now until the end of the year, erosion will strip layer after layer, in the process creating a Grand Canyon. The year-end visitors who gaze into this chasm will sense an impression of timelessness. In reality, however, this major geologic excavation is a recent happening.

At long last, Earth is being populated by those creatures with which we are most familiar and to which we are most closely related. By December 28 the Age of Mammals is in full bloom. Rodents are the most numerous, therefore most successful, as they will be until year's end. There is a great proliferation of primates, including among them the lemur. This small tree dweller has a short snout and large, widely spaced eyes providing stereoscopic vision. Agility, coordination, dexterity, even some intuitive understanding of gravity, are central to its arboreal survival, for a single fall can be fatal. Natural selection will therefore work to finely hone these skills, all of which require significant advances in evolution of the brain. Later, we who branch off from this line of descent will be deeply indebted to the time our ancestors spent in trees. Among the ground dwellers are the odd-toed ungulates—ancestral horses, rhino, and tapir.

We who appear last on this imaginary calendar have little doubt that mammals are superior to other creatures. But perhaps we are chauvinists, for if other fauna were less successful, then they need not exist. There is clearly not only room for a variety of organisms but a necessity for them. Nevertheless, there are certain adaptations, warm-bloodedness being one of them, we mammals successfully utilize. Our high metabolism necessitates a large fuel

FIGURE 42
Mammalian young benefit from long periods of instruction.

intake. Since we need oxygen to burn this food, we have developed a secondary palate separating mouth from nasal passages and enabling us to breathe and eat simultaneously. We can therefore take time for chewing, thereby speeding digestion so that nutrients are available quickly. In addition, because oxygen rapidly reaches our muscles, we have greater endurance than, say, a lizard. Then, too, our young are born alive and subsequently nursed, permitting a long period of instruction. Finally, all the selective pressure applied by our surroundings works toward the development of bigger and better brains, providing a greater capacity for learning and for applying our lessons profitably.

On December 30 seals, walrus, and sea lions suddenly appear. Soon they are joined by the even-toed ungulates—deer, cattle, pronghorn, pigs, camels, hippos, musk-oxen, goats, and sheep. It is evidently a busy day. For the first time, animals housing brains even vaguely resembling human ones are present. One, Proconsul, is apelike; its contemporary, Ramapithecus, however, may be ancestral to us. Its teeth are more level and human than those of its relatives. Furthermore, its face is much flatter, and its eyes are fully forward in stereoscopic position. Part of its existence is spent on the ground. The tall grasses there favor an occasional upright stance so that predators and prey can be spotted.

The Last Day of the Year. To us humans, whose vanity seems to require a status of dominance, the biggest surprise of the cosmic calendar has to be the lateness of our entry upon it. After all, 364/365 of Earth's history has transpired without us. That there is no evidence suggesting the biosphere functioned poorly in our absence is, at the least, humbling. To be sure, when we do appear, we will not go unnoticed.

Sometime in the forenoon of December 31, humans and apes launch separate lines of descent. But our apelike ancestors had acquired some useful traits that inheritance bequeathed to us. As an instance, one of the Australopithecan apes diversified from its herbivorous relatives to become an omnivore, perhaps because to do otherwise was to forfeit abundant food sources. The meat sought had a high concentration of protein. Consequently, less bulk had to be consumed and less time devoted to the gathering of it. With ample time came the prospect of hunting only big game. But such hunts, to succeed, required a degree of cooperative behavior. Societal planning and organization began to supplant individual efforts. Communication, if only by hand signals at first, was a valued skill. Henceforth a competitive advantage would accrue to those descendants whose brains enlarged commensurately with the expanded opportunities of communal action.

FIGURE 43
Humans appear in late afternoon of year's last day.

Hunting in tall grasses biased development toward erect posture. Bipeds appear early in the last half of the day. Their hands are free for tasks besides support. In concert with the rapidly increasing mental capacity, biped hands are quickly put to work in the production of tools of stone and bone. By 4:00 P.M. these developments attain considerable sophistication and variety in the person of *Homo habilis*, the first true man. Precious knowledge about the fabrication and use of tools is transmitted down through the generations via educational institutions. For the first time, a form of life has gained some leverage over its environment and is accelerating its use of it.

By 9:00 P.M. another of our ancestors, *Homo erectus*, has come forth. He is the first to have an endocranial volume as large as ours. Using his formidable mental prowess, he harnesses the power of fire, becoming the first planeteer to put to work a power greater than his own. These last few hours of Earth's history will reveal an incessant expansion in the use of external energy sources.

The last hour of the year is even more frantic. Man, by now recognized as *Homo sapiens*, is still nomadic, a wandering hunter and forager. His quest for food has spread his numbers worldwide. At 11:57 his communications acquire permanence: he utilizes art—cave paintings and the like—to put his inner feelings on display. At 11:59 an event occurs which will radically transform his existence and presage an alteration of Planet Earth, in its final minute, more extensive than any occurring previously over intervals of months. The epochal event is man's invention of agriculture, the ability to cultivate crops and to domesticate animals. The bond between life and its setting is apparently broken. Rather than passively (even gratefully) accepting the offerings of his environment, man learns to manipulate it actively to suit his specifications. Gone is the need to seek greener pastures; permanent civilizations can be, and are, erected. A cultural infrastructure arises, within which ideas multiply and reinforce. The pace of change is quickened a hundred, a thousand, a million times over. Urbanization demands a higher standard of public health. Inventive man responds with community water and waste-disposal systems. The death rate plummets. Nothing, however, changes the rate of births. As a result, the human species becomes the first to threaten its survival by overburdening its habitat.

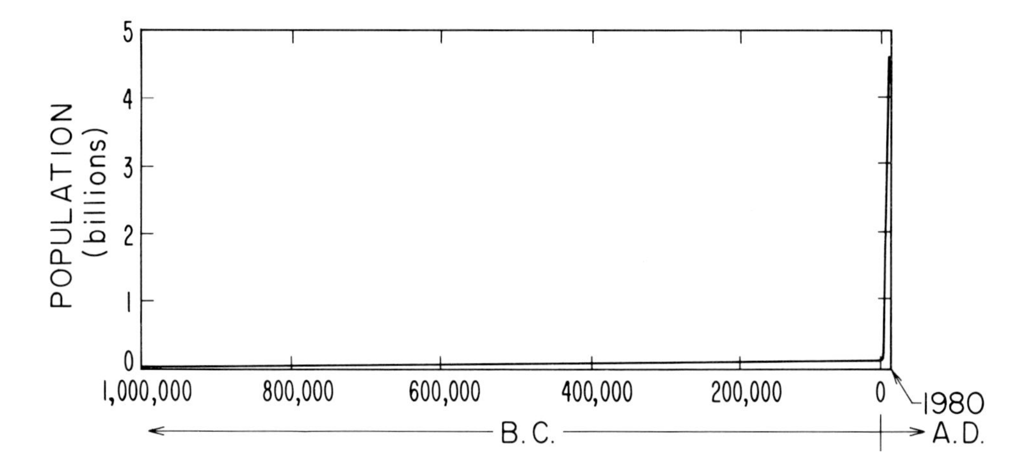

FIGURE 44
Human population. Can the ever more rapid rise in numbers continue indefinitely?

The wheel is invented at 11:59:25 P.M. At about the same time, a bristlecone pine tree (fig. 38) is born which will survive until the end of this story, our oldest living connection with the past. Christ is born at 11:59:45. One and a half seconds ago, 11:59:58.5, man launches another revolution, this one industrial. The machines with which he extracts power from nature stamp out an endless stream of ever more exotic products. Until year's end man will irreparably scar his planet seeking raw materials from it to feed his machines and then dumping his unwanted waste upon it. Twice in the last half second war encompasses the entire globe. An atomic bomb is exploded a fifth of a second before year's end. At 11:59:59.94 P.M., a blink of an eye ago, man becomes the first creature to leave his place of origin.

Suddenly it is now. A story has unfolded which envelops each of us in its magnificence: not as individuals endowed with special significance but as members of a family including all biota that have preceded, or will succeed, us. We are parts of a system vastly transcending us, only temporarily hosting the living presence while it continues its ongoing evolution.

Our ancestry predates the process herein detailed for Earth. Since the chemical substance of our being was inherited from stars, we may share kinship with other living inhabitants of the universe. If so, we should press our search for their existence, perhaps some far-distant day contributing to, and benefiting from, an active membership in a galactic community. Seemingly, the multiplication of cultural ideas within such a supercivilization would be proportionately as great as was the case when permanent civilizations of humans formed here on Earth. If, however, Earth hosts the sole biosystem in the universe, we perhaps bear the responsibility for propagating the life process outward from this Garden of Eden. Either way, although minute and temporary configurations of matter, we can appreciate the fortunate circumstance that some spark has endowed us with life. As a popular song has succinctly stated: "Life is rough. . . . Compared with what?"

FOR FURTHER READING

ALVAREZ, LUIS W.; ALVAREZ, WALTER; ASARO, FRANK; and MICHEL, HELEN V. 1980. Extraterrestrial cause for the Cretaceous-Tertiary extinction. *Science* 208:1095–1108.

ATTENBOROUGH, DAVID. 1979. *Life on Earth*. Boston: Little, Brown and Company. 319 pp.

BRONOWSKI, J. 1973. *The ascent of man*. Boston: Little, Brown and Company. Pp. 19–89, 259–319, 379–406.

CLOUD, PRESTON. 1978. *Cosmos, Earth, and man*. New Haven: Yale University Press. Pp. 143–268.

COLBERT, EDWIN H., ed. 1976. *Our continent*. Washington, D.C.: National Geographic Society. Pp. 18–162.

DAWKINS, RICHARD. 1976. *The selfish gene*. New York: Oxford University Press. Pp. 13–70.

GOULD, STEPHEN JAY. 1978 Feb. An early start. *Natural History* 87(2):10–21.

JACOB, FRANCOIS. 1977. Evolution and tinkering. *Science* 196:1161–1166.

MARGULIS, LYNN. 1971 Aug. Symbiosis and evolution. *Scientific American* 225(2): 48–57.

MAULDIN, W. PARKER. 1980. Population trends and prospects. *Science* 209: 148–157.

MAYR, ERNST, et al. 1978 Sept. Evolution. Nine related articles. *Scientific American* 239(3):46–230.

OPARIN, A. I. 1953. *The origin of life*. 2d ed. New York: Dover Publications, Inc. 270 pp.

SAGAN, CARL. 1977. *The dragons of Eden*. New York: Random House. Pp. 13–17, 81–104.

8
MAN AND THE EVOLUTION OF INTELLIGENCE

Doubtless all animals enjoy their existence; certainly we should wish it so. . . . But perhaps only to man, on this planet, is it given to enjoy with gratitude and understanding the sources of his delight. The two poles between which the universe evolves appear to be the immensity of space and the plasticity of matter, on the one hand, and appreciative, grateful minds on the other. Perhaps the latter, whatever the form of the body that supports them, are the goal or fulfillment of the whole stupendous process, with all the strife and pain which seem inseparable from the movement that gives birth to beauty and joy.

—ALEXANDER SKUTCH

Man's role in the evolution of the biosphere has been wholly disproportionate to the duration of his membership in it. Although our entry followed a natural path, our departure could be self-imposed. If we do opt for extinction, we shall join the many species whose disappearance we occasioned. Whether through human ignorance, neglect, or intent, the rate of extinction has accelerated from one species of plant or animal per millennium centuries ago, to one per decade by 1950, to one per year at present. What is this creature called *Homo sapiens*—man, the wise—and why is its influence so profound?

Anatomically it is unexceptional. Any competent paleontologist or anthropologist, given adequate skeletal remains, could identify them as human or not on the basis of posture and shape of backbone, length of legs relative to arms, location of thumbs, shape of jaw and details of teeth, and facial structure. Although in all these features we differ from fellow fauna, in none are we truly superior.

Where we do excel is in the fraction of our total weight which is devoted to brains. It is larger than among any other kind of animal. This is the fact that underlies the essence of our humanness and enables several nonanatomical features that lie at the source of human pride. For one, human behavior is considerably more flexible (less instinctive) in response to variable external stimuli than is that of any other animal. In addition, humans are capable of abstract thought and its expression, chiefly through spoken and written language. As a result, we can be imaginative and reflective, shaping our knowledge of past and present into visions of the future. In many instances, this permits foreseeing the consequences of certain actions and, therefore, of selecting in advance those whose outcomes are most beneficial. In its highest form this leads to a system of ethics or morality, to a sense of responsibility, and, at its pinnacle, to altruistic behavior. Humans also distinguish themselves from other earthly biota by the extent of their curiosity, the duration of their concentration, the capacity of their memory, and the mastery of their imitation. Add to this man's ability to make and use a variety of tools, thereby

FIGURE 45
Albert Einstein. Humans are capable of abstract thought and its expression. (Courtesy of the Archives, California Institute of Technology)

enabling him to perform tasks of increasing intricacy. A final distinguishing characteristic is man's social gregariousness, his uniquely complex banding into societies and cultures devoted to the common good.

Although many of these properties are shared with other animals, none is practiced to the degree demonstrated by humans. All are attributable to the exceptional percentage of our being which resides in our brains, for the

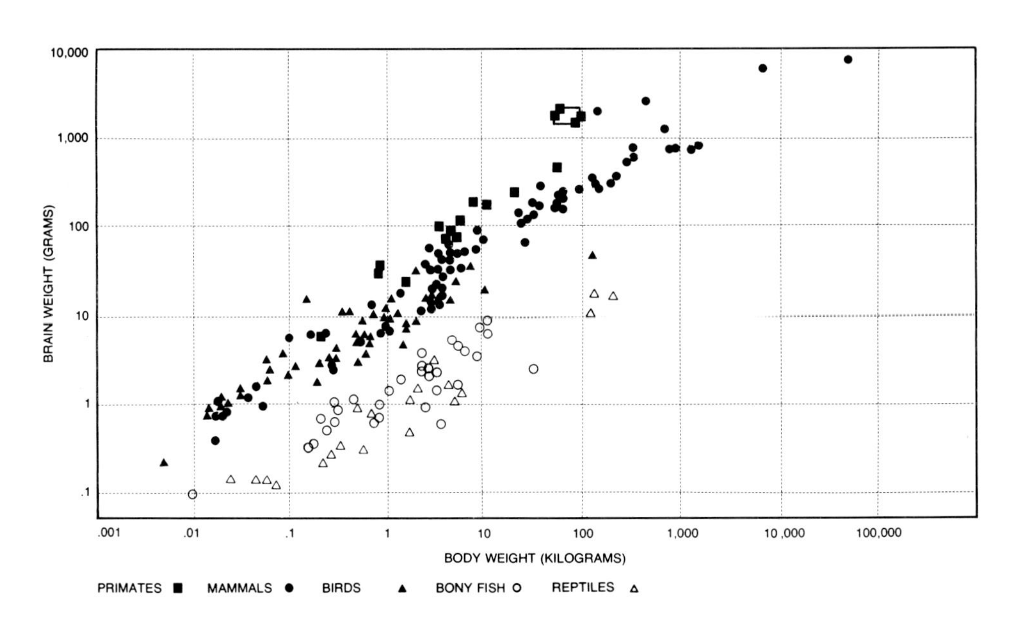

FIGURE 46
Brain weight vs. body weight. Open symbols for fish and reptiles, filled symbols for birds and mammals. Range of measurements for humans shown by four points connected by rectangle. Immediately to its right, at about same brain weight but slightly larger body weight (hence slightly smaller fractional brain weight), is dolphin point. (Reprinted from Harry J. Jerison, *Scientific American* 234(1):90–101, Jan. 1976. Copyright 1976 by *Scientific American*.)

greater capacity there has permitted the processing and storing of a larger flux of information than is possible among any species who are not similarly endowed. If the only storage location had been in our genes, our behavior would differ only slightly from that of reptiles; a sluggish, instinctive, preprogrammed response to externalities would have severely restricted our options.

The relationship between size of brain and size of body is best displayed graphically (fig. 46). Clearly the two sizes bear the same general relationship for fish and reptiles as for birds and mammals, but the encephalization of the latter species exceeds by an order of magnitude that of the former. Yet, since all four vertebrates exist at present, encephalization is not the only criterion of suitability. This suggests that the evolution of intelligence need not be inevitable in every bio-setting; survival in the rawest biological sense evidently does not demand it.

Nonetheless, one cannot dismiss the advantages of greater encephalization. Studies of fossil brains among direct descendants reveal a progressive increase and diversification in sizes relative to body sizes (fig. 47): that is, although small-brained species persist, both the average relative brain size and the spread of sizes increase with time. This indicates that more ecological niches can be successfully exploited as the brain becomes a greater fraction of the weight of a species, while, at the same time, stable niches can be maintained. The evolution of intelligence would therefore seem to depend upon the dynamism of the ecosystem considered, higher intelligence being the natural response to the multiple selective pressures of rapidly changing environments. To the extent that the emergence of life itself becomes a significant contributor to the dynamism of environmental change, biosystems containing some form of highly intelligent life may eventually be a reasonable fraction of the total.

The rate of evolution of encephalization among different vertebrate groups demonstrates just how quickly a new opportunity is seized. The standard pattern (fig. 48) is for a rapid increase in relative brain size immediately after the branching of a new order off the evolutionary tree, followed by maintenance at a steady value thereafter. Advanced intelligence (if that is indeed what is measured by the ratio of brain mass to body mass) originated with the mammals. It is worthwhile recalling how recently, in a geologic time frame, that development occurred. Even more dramatic is the

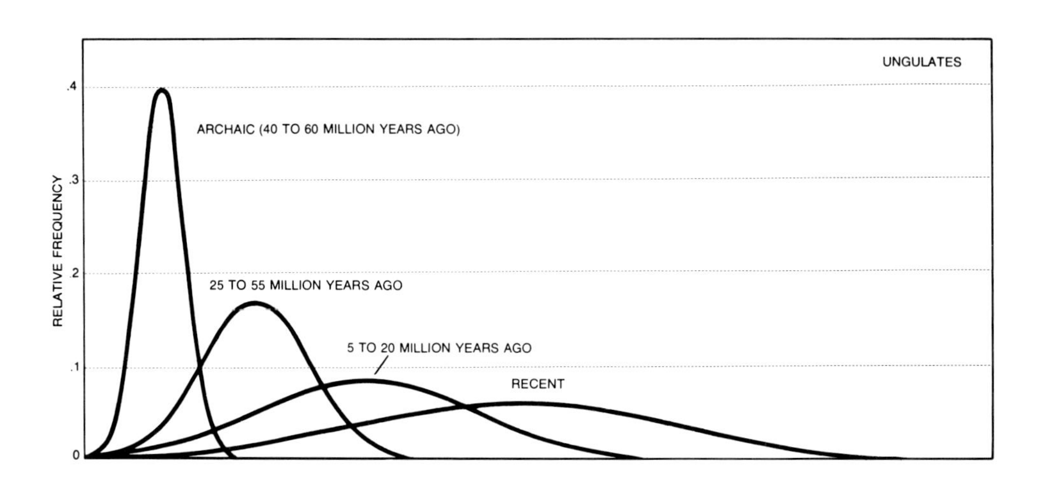

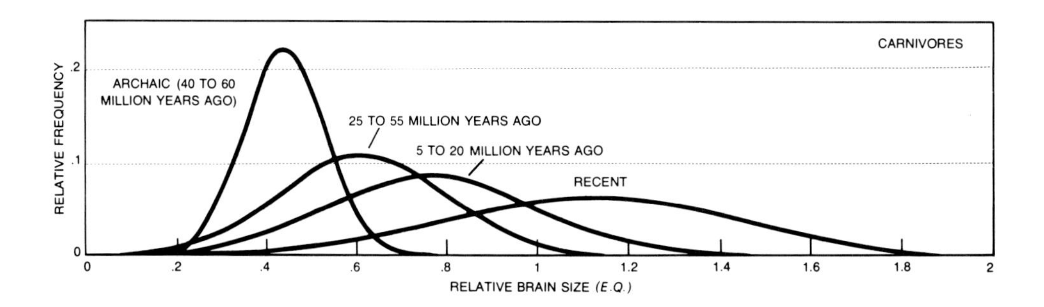

FIGURE 47
Histories of brain-size distributions among ungulates (top) and carnivores (bottom). For both, average brain size increases with time, as does spread of sizes. Small-brained species persist, however. (Reprinted from Harry J. Jerison, *Scientific American* 234(1):90–101, Jan. 1976. Copyright 1976 by *Scientific American*.)

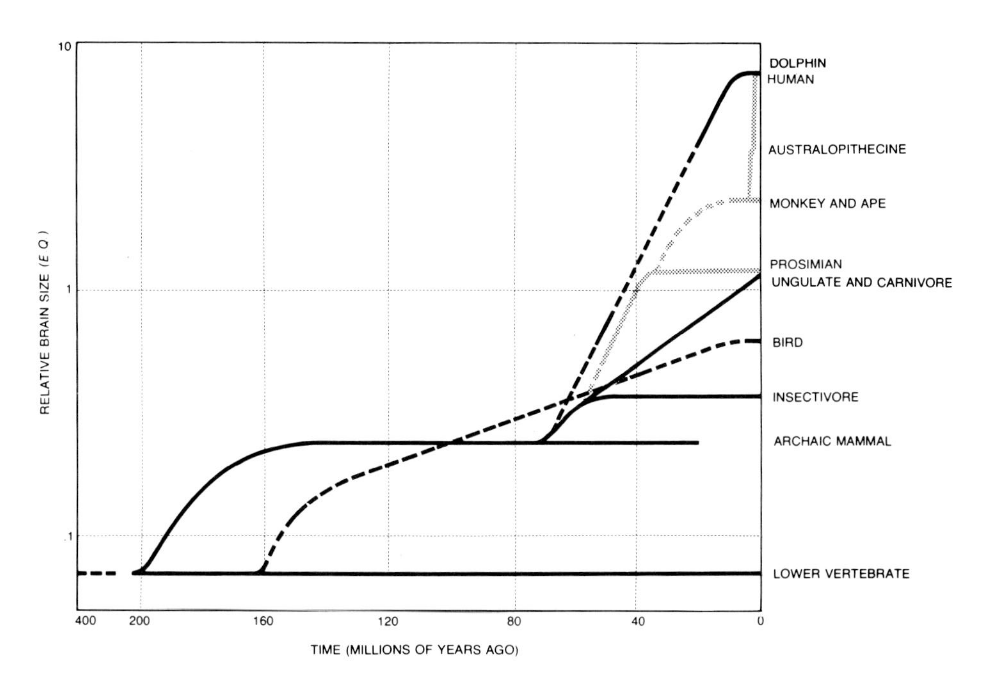

FIGURE 48
Rate of evolution of relative brain size typically exhibits rapid initial rise, then maintenance at constant value. Note recent and rapid increase of hominid brain sizes, necessary to reach level dolphins acquired ten million years earlier. (Reprinted from Harry J. Jerison, *Scientific American* 234(1):90–101, Jan. 1976. Copyright 1976 by *Scientific American*.)

realization that a good three-fourths of the mammalian era transpired at a fairly constant degree of encephalization before the extremely recent rapid increase originated (about four days ago on the compressed year-long calendar). The most extraordinarily rapid advance of all occurred among hominids, who, it should be carefully noted, needed a strong "finishing kick" to surpass the level of encephalization dolphins had earlier ascended to.

From this, and with the aid of considerable speculative license, we can infer two messages. First, many ecological niches are easier to occupy than the ones requiring advanced encephalization. Apparently the genes contain all the basic instructions for the successful operation of living systems. Additional information collection, storage, interpretation, and retrieval can be considered luxuries from a biological perspective, to be invoked in a biosystem only after the easier niches are fully exploited. It follows that of all the life-bearing communities beyond Earth (if any), only the subset of mature ones could host intelligent life forms.

A second observation is that Earth's biosphere has contained intelligent life, defined as biota storing more information in brains than in genes, for only a small fraction of its existence. Since time is the ultimate arbiter of which life forms are, in Darwin's sense, "the fittest" for survival, it is too early to know whether intelligence is an advantageous adaptation. The burden of proof rests on humans, since our actions alone represent the products of a collective intelligence synthesizing the inputs from both contemporaries and predecessors. The multiplicative effects of this synthesis enhance our powers mightily. Our actions now have consequences of global extent. Since we have not yet accumulated, from experience, wisdom in utilizing our global powers, we should seek every possible source of assistance. While it is romantic and fashionable to propose searches for extraterrestrial intelligent life—proposals whose merit is acknowledged—a possible source of assistance lies nearer at hand. The dolphins and related cetaceans have survived with an intelligence close to our own for a much longer time than have we. Perhaps we should be their students; certainly we should be their conservators, not their exterminators.

The whole of the foregoing discussion is flattering to human egos. What creature would not boast about possessing the highest brain-to-body ratio? Perhaps, however, we overexalt our own worthiness. Before self-indulging too extravagantly we should critically examine our biological role. The most all-encompassing perspective for doing so adopts the "Gaia hypothesis."

Gaia was a Greek goddess in whose body all living things were organs. She gave birth mythologically to the first race of gods and, later, to the human race. After fading from prominence for several centuries, she enjoyed a reincarnation inspired by environmental activists of the 1970s, this time as Mother Earth. The hypothesis (and at this time it is certainly no more than that) to which she is assigned asserts that life, acting as a collectivity and over a global basis, actively regulates and modulates the environment in just such a way as to optimize the very conditions under which that life can flourish. Earth's biosphere (life) is not independent of the atmosphere (air), hydrosphere (oceans), or lithosphere (soil). Instead, all are parts of a coherent whole. Insofar as that whole maintains a constant temperature and a compatible chemical composition—in short, a benign homeostasis—within a constantly changing setting, it can be considered alive.

Is it conceivable that life can modify for its own benefit, say, the composition of the atmosphere, or the salinity of the sea? The answer is affirmative, and the evidence lies in various parameters of Earth's environment which are anomalous in comparison with those of its nearest neighboring planets. Both Venus and Mars have approximately the same size as Earth, and all three were formed simultaneously from similar portions of the protosolar nebula. Differences in the fluxes of solar radiation they receive, consequences of their different orbits, can readily be accounted for. One therefore expects an interpolation between the properties of the planets adjacent on either side to approximate conditions on Earth. This expectation is not realized, however—not by a wide margin.

Earth's atmosphere has "too much" oxygen and "too much" nitrogen; hundreds of times more than a Venus-Mars interpolation would suggest. Likewise, there is "too little" carbon dioxide and sulfuric acid. In addition, geologic and fossil evidence both indicate that the temperature of the planet

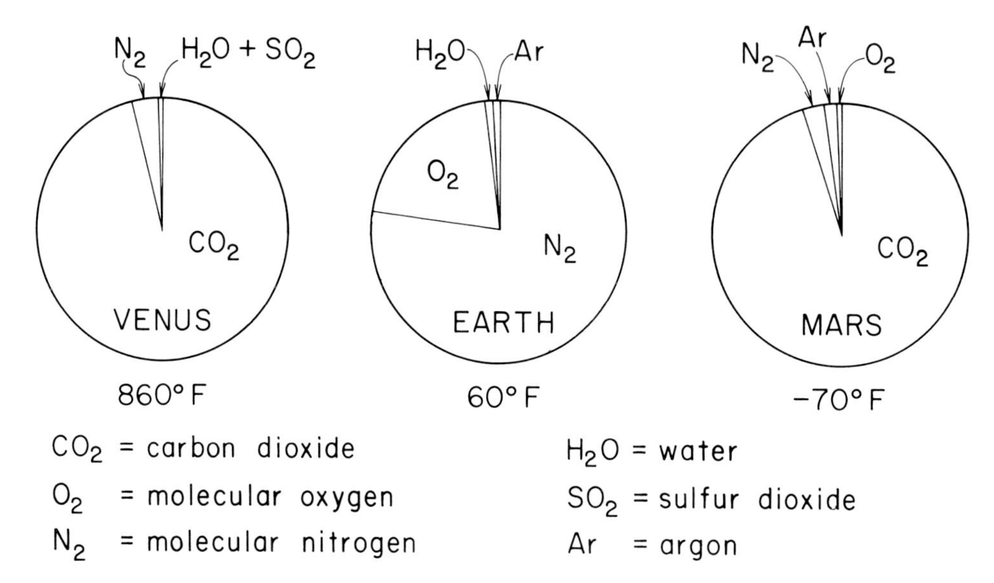

FIGURE 49
Atmospheric compositions and surface temperatures of Mars, Earth, Venus. Is it mere coincidence that Earth, teeming with life, departs from interpolation between its adjacent neighbors?

has changed little for the past few billion years despite a possible doubling or trebling in the sun's output during that time and despite major changes in the chemical composition of the atmosphere. Furthermore, the soil "should be" sterile; for with so high an atmospheric oxygen content, acid should be raining from the sky. Even ignoring this happenstance, ordinary rainfall, aeons of it, "should have" washed chemicals vital to life processes into the seas.

The mere presence of a fluid hydrosphere is another anomaly. If, for example, at any time in the past the temperature of the planet, on the one hand, had fallen slightly, some liquid water would have frozen into glistening snow and ice. As a result, more of the incident solar radiation would have been reflected, driving the temperature farther downward. Eventually a frozen Martian landscape would have resulted. On the other hand, had Earth's temperature risen, additional evaporation from the seas would have enhanced the cloud cover. More heat would have been trapped by the thicker blanket, and the warming cycle would have run away in the opposite direction toward a hot, smoggy Venusian environment. Informed judgment predicts a terrestrial environment that is either gaseous or solid. Yet liquids are what dominate Earth's surface.

Can it be mere coincidence that the "anomalous" planet happens to be the one teeming with life? Or has life seized control of the operation of Earth's "metabolic processes"? After all, the deviations from expectation are in every case in the sense to benefit life. And every deviation represents an instability, a disequilibrium. The systems most exemplary at maintaining states of disequilibrium with their surroundings are the living ones. A dead and fallen tree gradually decomposes and ultimately merges into the soil of its birth, thereby achieving an equilibrium condition. But throughout its lifetime, it defied the high entropic condition that was its destiny by drawing from its environment the nutrients and energy necessary to power an unstable, almost contrapuntal, existence. To the extent that Planet Earth, regarded as a single, interconnected entity, sustains a defiant posture of disequilibrium, it, too, can be regarded as living.

It is instructive to examine how Gaia maintains a global life-support system. For example, what has kept the oceans' acidity at a nearly constant level for the past three-and-a-half billion years, despite the nonstop production of various acids by ubiquitous plant life? The answer can only be seen with the aid of a microscope. Those primitive, single-celled microbes who first inhabited Earth play a crucial role in the decomposition of once living matter. While breaking it down into constituent elements, they release a wide range of gaseous products to the atmosphere. Among these is ammonia, a substance capable of neutralizing acids. Nature has (fortuitously?) provided a mechanism for preventing excesses threatening to life's continued existence. Note that the mechanism, microbial protozoans, is not among those biota we humans celebrate as heroic: no kings of the jungle there, not from our anthropocentric vantage.

What, then, of the oxygen content of the atmosphere? Absent initially but subsequently introduced by photosynthetic life, oxygen must now be present at its current level or devastating consequences will follow. On the one hand, should the level rise, spontaneous combustion will ensue, searing the plant life off the face of the globe. On the other hand, a fall in oxygen level would asphyxiate most animals. How are both unpleasant extremes avoided? An interactive ménage à trois protects the safety of all.

First of the trio are photosynthesizing algae, ocean residents that are oxygen producers. The oxygen consumers are tiny oceanic animals called zooplankton. A large population of these animals means a thorough utilization of the algae-expired oxygen, leaving little for escape to the atmosphere. Conversely, few zooplankton means incomplete interception of oxygen on its way from algae to atmosphere. With only these two participants, the system might be unstable: a spurt in the population of zooplankton, for instance, might unleash the savage consequences of atmospheric oxygen depletion. Fortunately, a third apex of the triangle regulates the numbers of zooplankton and thereby indirectly controls the flow of oxygen to the atmosphere. The critically important governors are marine bacteria.

As man does not live by bread alone, neither do zooplankton live by oxygen alone. Among their other needs are the nutrient nitrogen. The bacteria, however, likewise compete for this precious resource. The ferocity of the competition depends upon the quantity of oxygen in the environment. If the bacteria sense ample oxygen, their appetite for nitrogen subsides. The zooplankton then prosper and proliferate. As a result, the flow of oxygen to the atmosphere is reduced—but not dangerously so, for when the bacteria sense

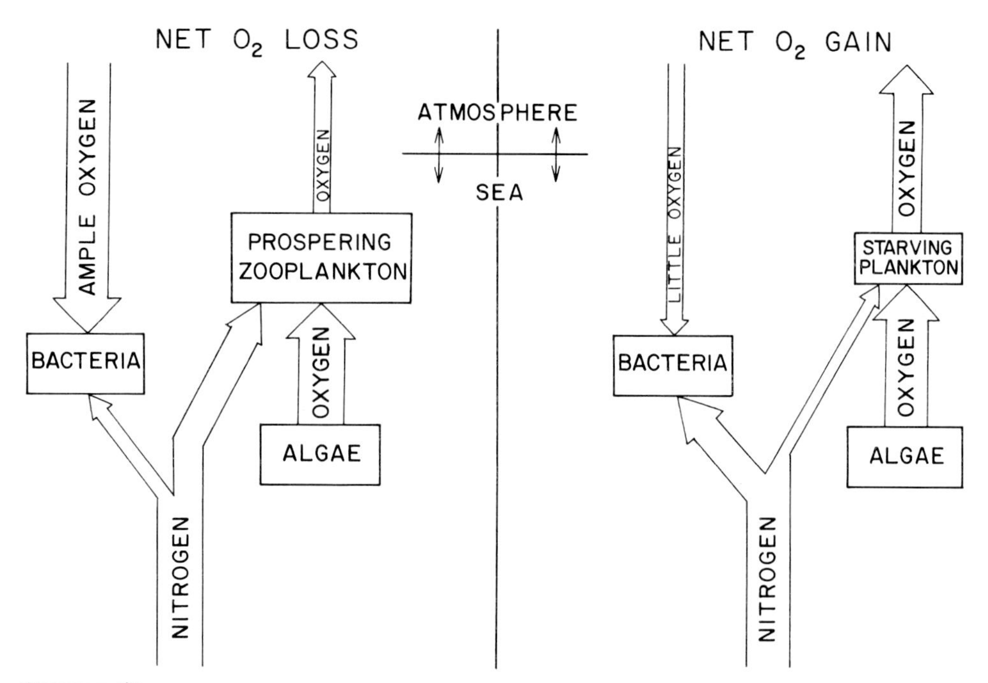

FIGURE 50
Regulation of atmospheric oxygen. At left, ample oxygen for marine bacteria lessens their demand for nitrogen; hence zooplankton population increases, and little oxygen passes from algae to atmosphere. At right, deficient supply of oxygen causes bacteria to consume nitrogen; plankton suffer, and little oxygen from algae is intercepted on way to atmosphere.

an atmospheric oxygen deficiency, they greedily consume nitrogen. Zooplankton starve, their numbers plummet, and oxygen flows more abundantly to the atmosphere. To the successful operation of this intricate interplay among microscopic coplaneteers we owe our existence.

Additional examples abound, but each emphasizes the prowess of these tiny creatures so distant from us in the evolutionary progression. Yet it is we humans, supported by the huge warehouse of extragenetic information stored in our bloated brains, who have arbitrarily divided life forms into categories of "higher" (e.g., mammals, especially primates, and birds) and "lower" (bacteria, algae, and lichens). If, however, the Gaia hypothesis is correct, the "lower" forms are the ones essential to the maintenance and operation of the global life-support apparatus. The higher forms are almost parasitic, profiting from the industry and enterprise of their primitive ancestors while contributing nothing of essence to their efforts. Perhaps neither form should malign the other. In reality, we are all woven into the same web.

We can now see the paradox of the human condition. Biologically we are superfluous. The Gaian overseer of the planet could easily adjust for our absence. In another sense, however, we may be essential. By virtue of our collective accumulation of knowledge, we can reconstruct the events that led to our existence as well as appreciate the spatial and temporal framework within which that existence takes place. We can even see that our degree of insight and ordering need not be uniquely the province of this planet. If, however, we or some other form of advanced intelligence had never appeared, how would the universe have been known to be? Or is it merely chauvinistic to wonder?

Evolutionary heirs to the marvels of the mind have contracted a Faustian bargain. In exchange for insights into their origins, status, and purpose, they are saddled with heavy burdens of responsibility. The same knowledge that permits enlightenment also conveys power. The two, knowledge and power, rapidly accelerate their expansion, each increment hastening the implementation of the next. The result is the acquisition of global powers by some species before its members are aware of their might.

A dangerous transition from passive recipient of a given environment's bounty to active molder of it must be made in an exceedingly short time. That time is now for the human inhabitants of Planet Earth. Our future, and that of our coinhabitants on the planet, is what we will it to be. To err now would be to end a glorious experiment in living at precisely the moment it is poised to sweep the universe into a fourth major era. The three preceding ones have been dominated by, in order, radiation, then matter, and finally life. We, here and now, are ready to enter an era dominated by cognition. Surely our curiosity compels us to sample its rewards.

FOR FURTHER READING

CLOUD, PRESTON. 1978. *Cosmos, Earth, and man*. New Haven: Yale University Press. Pp. 247–268.

COUSTEAU, JACQUES-YVES. 1977 Jan. 8. Can't find water in the river. *Saturday Review*. Pp. 61–62.

HAPGOOD, FRED. 1977 Dec. Gaia: the harmony of our sphere. *The Atlantic Monthly* 240(6):100–104.

HOLLOWAY, RALPH L. 1974 July. The casts of fossil hominid brains. *Scientific American* 231(1):106–115.

JERISON, HARRY J. 1976 Jan. Paleoneurology and the evolution of mind. *Scientific American* 234(1):90–101.

JOHANSON, DONALD C., and EDEY, MAITLAND A. 1981. *Lucy*. New York: Simon and Schuster. 409 pp.

LEAKEY, RICHARD E., and LEWIN, ROGER. 1977. *Origins*. New York: E. P. Dutton. 264 pp.

LOVELOCK, J. E. 1980. *Gaia: a new look at life on Earth*. Oxford: Oxford University Press. 157 pp.

SAGAN, CARL. 1977. *The dragons of Eden*. New York: Random House, Inc. Pp. 21–47.

VERSCHUUR, GERRIT L. 1977 Dec. The Gaia hypothesis—our living planet. *Griffith Observer* 41(12):2–9.

9
CULTURAL EVOLUTION

There is nothing at all absurd about the human condition. We matter. It seems to me a good guess . . . that we may be engaged in the formation of something like a mind for the life of this planet. If this is so, we are still at the most primitive stage, . . . but infinitely capacitated for the future. . . . It is remarkable that we've come as far as we have in so short a period.

—Lewis Thomas

Humans entered this world as natural products of aeons of biological evolution, distant lineal descendants from the original unicellular progenitor. The history of that biological evolution was episodic, with occasional bursts of rapid change punctuating longer intervals of stability. In the mean, however, the unfolding of life was slow. The advent of mankind dramatically accelerated the pace of change. Although human claims to have invented a new type of evolution are extravagantly anthropocentric, man did exploit more completely than any of his predecessors the advantages of cultural evolution.

By cultural evolution we mean the transmission of knowledge accumulated over several generations. There is perhaps some upper limit to the amount of information that can be transferred accurately via genes. Too large a set of genetic instructions risks too high a mutation rate. After a relatively few generations, the original instructions may become scrambled beyond recognition. But this fate is spared information that is transferred extragenetically. A large and accurate body of information, guiding the knowledge and behavior of individuals, can accumulate over the course of hundreds or thousands of generations, each generation perhaps contributing only minute advances. No longer must each generation discover every insight anew. Instead, each starts with a base to which it can only add. The result is an explosive growth of knowledge.

As an example, some remote ancestor of ours may have been able to associate weather variations, especially the changing of the seasons, with the arrival of game, enhancing his success at hunting. Such acquired associations are exactly what cannot be passed to descendants using the genetic code. But given some skills at communication, they can easily be shared among contemporaries and, subsequently, successors. The latter may further hone the hunting skills of their societies by adding insights of their own. Eventually, of course, some generation may so tire of the chase that they instead capture and domesticate their former prey. They can do so only when their knowledge of animal behavior permits. That is, this relatively modern man is probably no more intelligent than his Cro-Magnon ancestor; he is more powerful only because he has access to a greater store of information. Nor are we today genetically better endowed than the first animal husbandmen of ten thousand years ago. But our productive modern farms symbolize the advances made possible by distilling and integrating the incremental improvements instituted by the hundreds of intervening generations.

Of course, cultural evolution was not totally divorced from biological. While acquired traits could not be transmitted genetically, advantages did accrue to those most facile at absorbing the lessons of yesteryear. The rewards deriving from skill at exploiting available knowledge were so great that they created a selective pressure toward ever greater encephalization. Over time, *Homo sapiens* evolved into *Homo sapiens sapiens*.

For this, however, a price was paid: specifically, the slowness of development of human children. No other species has such helpless offspring for so long a time. Years, if not decades, are required for the provision of adequate training and education. The other side to this is that the character, skills, and behavior of the adult individual are determined as much, or more, by this directed education as by inborn capabilities. As a result, the possibilities for flexible behavior are endless.

Cultural evolution would not have been so rapid had it only enhanced individual skills. Fortunately, the communication grid over which training and education were distributed also tied man to his fellow man. A collective intelligence evolved whose total effect vastly exceeded the sum of its parts. Partly this is because more and better ideas are generated when their rudiments are bounced among several fertile minds. Just as, in a single mind, the number of interconnections, or synapses, between neurons determines that mind's power, so too may more minds in contact strengthen the collective intellect. A second advantage of collective behavior is that it permits specialization of function, not as a strictly genetic trait but as a product of society's training so that the individual can better serve the group. This means, of course, that hierarchies of command must develop apace.

Both specialization and chains of command were evident in man's earliest group actions, namely, defense from predators and acquisition of food. Over the many generations since, huge, complex organizations have evolved whose operation defies the understanding of any single member. They seem to resemble living organisms in the intricacy of their multiple interacting feedback loops.

Consider, for instance, a space mission. The satellite is built in one aerospace organization, its launching rocket in another. None of the people who physically build these structures—the welders, painters, electricians, fabricators, and so on—knows how to build in its entirety the very system which, together, they actually do construct. Likewise, the designers who guide their efforts probably could not perform all the necessary assembly steps. Elsewhere, some scientists plan the satellite's trajectory, while others design the experiments to be performed. When the experimental data are transmitted from the target to Earth, they pass through a network of telescopes linked to one another and to a command center by telephone, telegraph, and microwaves. None of the operators of the network fully comprehends all aspects of the space mission. Each performs a narrowly specialized function—for example, clock synchronization—without which the mission would fail but which, by itself, is but a minor part of the overall effort. The individual members of the team are of no consequence, either: since a mission requires a decade or two from inception to completion, those involved at its start may not be present at its end. Clearly an organization has arisen whose collective ability dwarfs the contributions of any single individual and which, in fact, operates smoothly despite the absence of any individual with detailed knowledge of all its inner workings. Collaboration, cooperation, coordination, and communication give man a collective power that threatens, in some cases, to overwhelm his control.

The foregoing example illustrates how successfully mankind, in its cultural evolution, learned to translate the knowledge it accumulated into the power it exercised. From the earliest primitive tools, such as clubs and stone flakes, to the most recent products of his technological cornucopia—computers and guided missiles, for instance—man has rapidly extended the physical capability of his body and harnessed external energy sources to his own purposes. Some inventions, the microscope and telescope among them, extended the range of his senses. Others, like the discovery of metallurgy, broadened his options for molding nature's bounty into useful structures. Still others, his vehicles of transport, gave him the mobility to spread more extensively than any other species on the planet and even, most recently, to leave it altogether. And all of these astonishing developments, it is worth reemphasizing, occurred in the last ten-thousandth of a percent of Earth's history!

A gigantic leap taken by humans a mere six thousand years ago (forty-two seconds if all of Earth's history occupies one year) which elevated them to a new plateau of technological prowess was the use of symbols (fig. 51). Those understandings and associations that had formerly been held internally, whether genetic or extragenetic, could now be made external by using letters and numerals. Greater fidelity of information transmission was assured, as was extent of distribution. In addition, quantification enabled immense refinements in the precision of tool manufacturing. In his writings, his pictures, and, later, his recordings, both aural and visual, man possessed a store of information unique to his kind alone. His powers were magnified accordingly.

As cultural evolution gave humans progressively greater control over their environment and more complete utilization of its resources, it became

FIGURE 51
Petroglyphs, Dinosaur National Monument. Cultural evolution accelerated dramatically when humans mastered symbols, displays of information having permanence and fidelity.

possible for larger groups to support themselves in relative stability. The human history of the last ten thousand years has seen man congregate in ever larger cities. None of his coplaneteers has ever banded into such immense units. We are far and away the most social of all animals. Yet we simultaneously enjoy the broadest geographic distribution.

Cultural evolution has endowed humans with a unique status on the planet. Now we must determine whether events can be controlled. Are we the masters or the servants of our technologies? To answer, we must consider some of the unforeseen effects that have accompanied our explosive technological development.*

We begin by noting the extent to which technology has attenuated experience. Life still "roller coasters" over peaks and valleys, but neither extreme is as pronounced as it was even a century ago. Some of the excitement of living has been flattened. Where, for instance, is our sense of the miraculous? When man first ventured onto the moon, the entire world's attention was riveted upon his every step. But what of the fifth and sixth Apollo missions? Who were the astronauts involved? Had the "impossible" become merely commonplace? Were we so satiated with technological wizardry that nothing could any longer stun anyone? There are even people who persist in believing that the lunar and other space missions were only staged productions. To them, at least, and to a lesser extent all the rest of us, technology has blurred the distinction between the real and the artificial.

Another change fostered by technological advance is the extent to which humans have divorced themselves from the rest of nature. The changing of the seasons is a natural process over which we exert no control. Its annual rhythm

*Not all the world's people have shared equally the advantages and disadvantages of technological development. The following discussion is addressed to those residents of so-called developed nations, the "haves" in a world that includes "have-nots."

can remind us of our limitations, but only if we are attuned to its variations. Instead, we have provided ourselves with central heating and air conditioning, with humidifiers and dehumidifiers, with buildings whose windows cannot be opened. The result? A climate that is the same everywhere at all times. Houston summers can be endured, as can Fairbanks winters. Space, geographic setting, has become homogenized, and time has lost the annual regularities that mark its progress. Even when shuttling between buildings we avoid contact with external realities by carrying the same artificial environment with us in our cars.

There was a time when the food on our tables specified both locale and season. But that was before frozen, canned, and dried foods preserved every product indefinitely and before transportation networks enabled its worldwide distribution. Today, Saudi Arabians dine on strawberries in midwinter. Those who grow the produce formerly endured annual extremes of flood and drought, until the technology of storage reservoirs and irrigation canals smoothed out nature's own dispensation of water. Weather modification may soon succeed in altering the cycle further. While all of the changes yield benefits, they also extract costs: by separating ourselves from nature we have dulled the feeling of what it is like to be alive and a part of it.

Technology is also responsible for erasing the distinction between indoors and outdoors. That this change has been rapid is well illustrated by the fact that some guests at a typical backyard barbecue will remember when people ate indoors and went to the bathroom outdoors. A major change has evidently occurred in less than a human lifetime. Today municipal water systems provide a bottomless supply of water to every residence, while waste-disposal systems transfer an unpleasant problem beyond the home. The ubiquity of glass in modern architecture likewise broke the barrier between in and out. Moreover, once portable radios, televisions, and tape recorders appeared, a person could carry the indoors with him wherever he went. The reverse was also true: domed stadia hosted outdoor games in indoor settings.

Previously, no mere mortal could be everywhere all the time. Events happened somewhere, once—then never again. The unrepeatability of each passing moment gave meaning to life, as well as to death. Technology, however, made experience repeatable, as witness, for example, the assassination of John F. Kennedy. How many of the author's generation of Americans were present in Dallas at the site of the president's death? But who among us cannot visualize the parade route, the fallen president, his desperate wife, and other details of this tragic event? The miracles of television, radio, phonographs, movies, books, magazines, and more of the same, bring every event everywhere, endlessly. In fact, one can live a completely vicarious existence, shunning such adventures as scaling mountains, fathoming oceans, or competing in sports, yet sharing the resultant sensations without leaving home. Again, the boundary between what is real and what make-believe has been blurred.

All the gadgetry mentioned, one could argue, did make life richer. But when everyone, everywhere, had the same devices, their lives may instead have been impoverished by the diluted sensation with which they experienced living. Our cultural evolution has delivered to us a paradox: while attempting to enrich the lives of everyone, we may be accomplishing just the opposite. Have we, in fact, lost the meaning of an enriched life? Is it to be found in a higher standard of living, or in the thrill of pursuing one? Is it the result of having things, or in seeking to have still more? Is it in being well off, or in becoming better off? Some of these questions are never asked (let alone answered). We seldom have an opportunity to say whether this or that would be a good idea. Rather, happenings just seem to take place. Some intangible momentum gallops across our lives, propelling us to who-knows-where for who-knows-what reason. Can we, should we, reassert control?

The buildup of momentum began when we committed ourselves to the consumption of novel items by the greatest numbers of people. To provide such, we developed assembly-line technologies. More and more goods spewed forth in less and less time. Once an item had been sold to everyone, however, it had to be changed to be sold to each person again. Planned obsolescence, practiced most garishly by the automobile industry, became the norm. The incessant output of mass production became the driving force behind innovation. Yet the innovation had an eerie quality. It was so planned and predictable that it lost its power to surprise. Novelty became routine. And when it did, a form of social engineering became necessary. Advertising sought to require society to remold its needs so that the new products were useful to them. Devices were not made to satisfy people; people were made to be satisfied by the devices. When that reversal of logic was accepted, mankind forfeited control of its destiny.

Having ridden the wave of high technology to near omnipotence, we have fallen victim to the mercy of the unpredictable advance of human knowledge. Few creations of our hands and minds began with malicious intent. The innocent pursuit of knowledge of the living world about us has, however, led us to the brink of genetic engineering, a force of potential good and evil. But no biological pioneer foresaw this ability, or sought it if he did. Likewise, those curious early investigators of the structure of matter did not foresee the unleashing of nuclear fission (atomic energy). Yet both biologists and nuclear physicists—and these are but two examples among many—have forced upon us a new kind of decision making. Their every increment in understanding unveiled a new perspective from which the direction and magnitude of the next step were obvious. To fail to take that step was to waste the whole of their previous efforts. In this way decisions became negative only. Either let things happen, a nondecision, or say no, rendering futile the many steps that led to the decision of abandonment. How radically different this is from the positive pursuit of desirable goals.

Man finds himself with abilities he did not seek and powers he must not use. He is the victim of the speed of his own cultural evolution, which granted him extraordinary skills before he had acquired any experience in utilizing them. As a result, he has unleashed forces over which he exerts little control. He is curiously omnipotent and impotent at the same time. But he does not have the luxury of bemused contemplation of his dichotomous nature. He, alone among the creatures on this planet, must consciously and conscientiously select the course of his subsequent evolution. None of the decisions, or lack thereof, which preceded his current predicament are irreversible. Man can seize control of the events that envelop him. In fact, he cannot do otherwise.

FOR FURTHER READING

Bonner, John Tyler. 1980. *The evolution of culture in animals*. Princeton: Princeton University Press. 216 pp.

Boorstin, Daniel J. 1973. *The Americans, the democratic experience*. New York: Random House (Vintage Books). Pp. 305–408, 523–598.

Bronowski, J. 1973. *The ascent of man*. Boston: Little, Brown and Company. 448 pp.

Dubos, René. 1968. *So human an animal*. New York: Charles Scribner's Sons. 300 pp.

Ellul, Jacques. 1964. *The technological society*. New York: Random House (Vintage Books). 465 pp.

Galbraith, John Kenneth. 1958. *The affluent society*. Harmondsworth, England: Penguin Books. 298 pp.

McGhee, George C. 1976 Feb. 7. Education as evolution. *Saturday Review*. P. 5.

Thomas, Lewis. 1974. *The lives of a cell*. New York: Bantam Books, Inc. 180 pp.

———. 1979. *The medusa and the snail*. New York: The Viking Press. 175 pp.

10 UNIQUENESS OF EARTH'S PRESENT

Diversity may result in some loss of efficiency. It will certainly increase the variety of challenges, but the more important goal is to provide the many kinds of soil that will permit the germination of the seeds now dormant in man's nature.

—RENÉ DUBOS

Man is neither the measure of all things nor the source of all value. Terrestrially, he is not a biological necessity, but a luxury. He is not the end point of evolution but only its most recent product. He dare not dominate life on this planet but rather must coexist with it.

The planet on which man resides is in no way special, nor is it at the center of the universe. Its parent star, the galaxy in which that star exists, and the cluster to which that galaxy adheres are likewise cloaked in anonymity. Abundant likenesses of these structures are remote in distance but close in kinship: all share the same ancestry of cosmic evolution. Earth is therefore unlikely to be either the only locale where life exists or the only habitat where intelligence has arisen.

If intelligent life emerges, a new kind of evolution, cultural, supersedes its atomic, chemical, and biological predecessors. It differs from these predecessors by virtue of its speed of transformation. Supersonic aircraft, whose wingspans exceed the initial flight path of the Wright Brothers' Flyer, follow their flying pioneer by a mere fifty years, or less than a typical human lifetime. This illustrates how rapidly knowledge, and the ability to exploit it, accumulates. Inevitably this progression provokes a crisis: at the moment when the species driving the cultural evolution amasses powers global in extent, the survival of its entire biocommunity is endangered. Problems of a previously unencountered nature appear. Their solutions cannot be postponed. We humans of the present generation on Planet Earth are among the most recent to have crossed the evolutionary threshold to global powers.

When we consider how we arrived at our current predicament, we note striking similarities between cultural and biological evolution. The gene is the fundamental unit for transmitting information biologically. The special property that gives it so prominent a role is its self-copying ability: it is a replicator. Of course, no natural replicating process is perfect. Life, if not extinguished altogether, would be despairingly monotonous had fallibility not been built into the copying process. Instead, each error or mutation offered the opportunity for initiating new living forms, adaptable to new environmental niches. Nearly all of these naturally occurring experiments were failures; otherwise life would have careened between unicorns and mermaids with such rapidity that the stability essential to its survival would never have arisen. But the small fraction of successful mutations eventually enriched the biosphere with biota as diverse as warthogs and prickly pear cacti. The reason why life continually changed was that the various replicating entities had different fates when pitted against their environments. A differential in ability to survive soon showed up as a shift in relative numbers of competing species.

Are there analogies in the case of cultural evolution? There are indeed ideas, thoughts, inventions, artistic styles, musical themes, fashions of dress, idioms of expression, architectural standards, computer algorithms, and other units of information whose propagation determines the pace and direction of a culture's changes. These units are the analogues of genes. They are transmitted, not as is the gene by a union of two contributing parents, but directly from brain to brain. In that sense our brains are almost public organs, openly exchanging information through conversations, electronic communications, printed media, hand-distributed leaflets, art works, photographs, billboards, and innumerable other conduits. All of us are enmeshed in a circuitry girdling Earth over which information reverberates with astonishing speed. As a result, the products of our minds are syntheses of the thoughts of others. We assemble ideas from bits and pieces gleaned elsewhere.

In cultural evolution the transmitted units of information are not so much reproduced as imitated. Clothing fashions, catchy tunes, and clever inventions acquire a permanence when they are widely adopted. Richard Dawkins, author of *The Selfish Gene*, has introduced the name *meme*, a contraction of the Greek *mimeme* (imitation), for these and similar examples of cultural phenomena that have been emulated in so many settings and for so long a time that they have become customs and traditions.

Not all of the imitations will be perfect. As in biological evolution, this fallibility is necessary to perpetuate the process of change. In the case of the gene, the imperfect copies were called mutations. With respect to the meme, the mutations are the bold, trend-setting innovations introduced by such true intellectual giants as Einstein, Picasso, Shakespeare, Mozart, and even Disney. Meme mutations occur every several generations in whatever field we consider. Without them the world would be a lot less civilized.

Let us also note how culture mimics a living process. Life evolves by the differential survival of replicating entities, culture by the differential survival of imitating entities. Ideas, thoughts, speculations, and dreams compete for the limited attention a given mind, or set of minds, can devote to any single topic. Those memes that survive this competition are the ones that excel in terms of longevity, fecundity, and fidelity of imitation. Over time, these crowd out the others.

Humankind is extraordinarily lucky to be a participant in both kinds of evolution. Although our genes may last forever, the exact collection of genes that is any one of us disappears rapidly. At each generation, our genetic contribution is halved. By contrast, someone who writes a book, designs a bridge, builds a better mousetrap, or composes a tune donates a contribution that may survive intact long after his biological self has revealed its mere mortality. Socrates' ideas still pervade our culture intact, whereas his genes are widely scattered among the huge number of progeny who have descended from him.

The memes we have inherited, plus those we originated, now empower us to breed ourselves out of existence, to annihilate the living presence on the planet, to pollute its whole environment, and to perturb the global climate. Capabilities so extensive in effect were unimaginable as recently as a mere century ago. A danger exists precisely because we may fail to recognize the uniqueness of this exact moment in Earth's history. In the overall context of cosmic evolution, during which many biosystems may already have encountered the threshold of global powers, it may not matter whether ours survives the crisis—as long as some do. But if *we* believe in the worth of the human phenomenon and truly value the good fortune that has proliferated life on our planet, it should matter to us. Faced with this unprecedented crisis of our own making, we can either cringe helplessly before the fear of catastrophe or accept the challenge as an opportunity to explore totally new evolutionary possibilities.

The greatest threat may not be biological extinction but, instead, an extinction of the cultural values that define our humanness. Unless we collectively summon the inner resources to develop the new values and institutions essential to a stable and cooperative worldwide civilization, we may produce a mental stagnation resulting from the oppressive regimentation and coercive restriction of freedoms which will be seen as necessary to cope with ever more trying circumstances. And mental stagnation spells the death of cultural evolution.

Nonetheless, there is hope! It lies in the replenishment of the meme pool. The overwhelming strength of the system biological evolution has brought forth is rooted in its profligate diversity. The gene pool is sufficiently rich to contain many different ways to make a living. The collective biosphere therefore has the resiliency to surmount environmental challenge—earthquakes, fires, volcanoes, glaciation, and the like. By analogy, if our culture is to proceed through its present crisis, it will be because the meme pool is equally robust, equally richly stocked with varied alternatives. The leaps and bounds so necessary to accelerate the pace of social, cultural, political, and economic adjustment will occur as do mutations in the course of the evolution of living systems—that is to say, randomly and occasionally—whenever the spark of genius is ignited. But they will come only if adequate raw material in the form of basic information is already stockpiled.

We must therefore encourage the expression and propagation of alternative and original viewpoints. We cannot restrict the boundaries within which what is called "legitimate discussion" takes place. To do so is to deprive ourselves of potential solutions to future problems. Yet that may be exactly the trap into which the world, including those countries that pride themselves on freedom of expression, is falling. The "movers and shakers" in our society—be they political figures, captains of the media, corporate executives, labor leaders, economists, or assorted other "villains" who most benefit from the system as it is—are limiting expression to a narrow range clustered about some nebulously defined center that is basically the status quo. But the status quo is definitely not appropriate for a transition era. It will not suffice merely to conquer the notion that someone is a threat because he proposes changes to a system that has worked well in the past; we must go the next step of actively encouraging the cultivation and free exchange of alternative ideas.

We must also avoid submerging cultures which we define as alien. Within them, if only we were to look, might lie some alternatives that would help smooth the way through our multiple crises. A Western, industrialized life-style is not appropriate for all the peoples of the world. Nor is an Eastern, an Asian, a Latin American, or that of any other single culture. All contribute to a social fabric that is the stronger because of the many strands from which it is woven.

Moreover, we must preserve a certain imperfectibility in the meme-propagation process. Our connected minds occasionally need the jolt of absolute unpredictability and total improbability, without which adjustment to the pace of change can be too stiflingly slow. Historically, the civilizations that have withdrawn into pleasure gardens dedicated to the preservation of the comfort and safety they have achieved end their development at that level and wither away in the smugness of their supposed satisfaction. The same will be true in the future, although the decay will be ever faster.

Finally, as individuals we must assert the strength of character, independence of mind, and boldness of conviction to create and express memes of our own. If all are takers from the meme pool, and none contributors to it, the pool quickly dries up. It may indeed be easier to succumb to the narcotic of mass culture and to avoid the strenuous intellectual adventure of stretching one's mind with a new idea, but this very ease of submission deceptively masks a path toward extinction.

Nothing short of a new kind of evolution is required. I described it in the beginning chapter as participatory and collective. There is no sideline to which we can retreat; there are no subgroups of our species which we can ignore. In biological evolution, the modus operandi is unbridled self-indulgence, ignoring coplaneteers of both present and future. By contrast, the new evolutionary era we are entering will be dominated by cognition. That means man must project the outcomes of his many optional actions, then choose the ones promising the best long-term survival possibilities for the biosphere. The future will not be selected naturally, but consciously. Complete randomness will yield to loose direction. Creatures who seem cosmically insignificant must now assume full responsibility for the continued prosperity of their habitat. We alone on this planet are capable of the mental discipline to defy the selfishness of our genes and to override the indoctrination of our memes.

Will we accelerate our social adjustments to the appropriate, albeit bewildering, pace? Or does the long, tangled evolutionary thread that led to our existence end here and now on this one small planet? We have surely arrived at a fork in the road. One path plunges down what is probably a well-worn cul-de-sac. The other leads toward a future of unlimited, hopeful expectation.

Saguaro trees in a lifetime scatter millions of seeds by wind and water across the desert, of which but two or three attain full development. Metaphorical seeds of life may likewise be scattered profusely throughout the universe, with few indeed able to evolve into mature cultures. But unlike the saguaro's fate, adolescent biocultures effect their own destinies. Let ours choose consciously to continue, in the hope we may blossom into a global, thence perhaps universal, society of as yet unimagined beneficence.

FOR FURTHER READING

CLOUD, PRESTON. 1978. *Cosmos, Earth and man*. New Haven: Yale University Press. Pp. 302–363.

COLINVAUX, PAUL. 1978. *Why big fierce animals are rare*. Princeton: Princeton University Press. Pp. 212–233.

DAWKINS, RICHARD. 1976. *The selfish gene*. New York: Oxford University Press. Pp. 203–215.

DUBOS, RENÉ. 1976. Symbiosis between the earth and humankind. *Science* 193: 459–462.

FALK, RICHARD A. 1972. *This endangered planet*. New York: Random House (Vintage Books). 496 pp.

KENNAN, GEORGE F. 1981 Jan. Cease this madness. *The Atlantic Monthly* 247(1): 25–28.

MCHARG, IAN L. 1969. *Design with nature*. New York: Doubleday & Company, Inc. (Natural History Press). 197 pp.

MORAL, ROGER DEL. 1981 May. Life returns to Mount St. Helens. *Natural History* 90(5):36–47.

SCHELL, JONATHAN. 1982. *The fate of the Earth*. New York: Alfred A. Knopf. 244 pp.

INDEX

Designer: Marvin R. Warshaw
Compositor: Trend Western
Printer: Halliday Lithograph
Binder: Halliday Lithograph
Text: Galliard
Display: Galliard